Does Social Media Have Limits?

"'The Social Dilemma' somaesthetics meets the 21st-century body electric. Camila Mozzini-Alister's *Does Social Media Have Limits?* is a deeply reflected and profoundly inspiring tale of traversing the distance between our rabbit holes of social media and the black holes of seeking a self in the universe of technical images. An indispensable experience for those looking to find the singularity or just looking to regain consciousness."

—Anke Finger, *Professor of German Studies, Media Studies, and Comparative Literary and Cultural Studies, University of Connecticut, USA*

"*Does Social Media Have Limits?* is a living testimony of an artist who gave her own life and body to her work. It is difficult to categorize Camila Mozzini-Alister's work since it is, at once a work of philosophy, biography, performing arts, media archaeology, science fiction, as well as an innovative way of bringing together mind and screen through the practices of mediation in social media and spiritual meditation.

One of the most interesting aspects of this book is how the author uses her own voice and experience to relate to any other voice in the same quest to becoming more than just a body. Grounded in a rigorous phenomenological exercise, she calls attention to our deep subconscious 'desire for omnipresence' and brings forth a crucial question for humanity: after all, what are the limits of our 'progress?'"

—Josepa López Poquet, *Professor of Fine Arts, Polytechnic University of Valencia, Spain*

"This book by Camila Mozzini-Alister is original from any perspective we approach it. The originality is revealed by the relationship that the author establishes between mediation and meditation, leading her to propose the term 'medi(t)ation.' Its originality is consolidated by the writing style, combining a refined study on the best philosophers of technological mediation with transformative personal experiences with meditation. Mozzini-Alister's text is a living proof that objectivity and subjectivity either interpenetrate and become fruitful or do nothing at all.

However, the striking originality of the essay does not imply, as the most ingrained traditionalists might think, any loss of quality. The combination of refined study with the assumption of subjectivity itself, therefore of the I-who-speaks, amplifies the quality of the hypotheses raised and the arguments that she defends. When the author sees, for example, 'in the act of continually touching screens the search for a common desire in meditation practices: the desire for limitlessness, or enlightenment,' she literally enlightens our reading, helping us to understand our everyday life gestures, in front of the screens that surround us, as part of the search for transcendence.

This is a rare work, worthy of being read and reread again and again."

—Gustavo Bernardo Krause, *Professor of Literary Theory, State University of Rio de Janeiro, Brazil*

"Camila Mozzini-Alister's work is a daring combination of traditional scholarship and personal experience. Her approach to digital life highlights the non-evident spiritual components and possibilities of our technological existence."

—Erick Felinto, *Professor of Media Studies, State University of Rio de Janeiro, Brazil*

"In this fascinating book I am reminded of the brilliant works of William Irwin Thompson, and his classic, *The Time Falling Bodies Take To Light.*

However, while Camila Mozzini-Alister is equally poetic and dazzles us with her stories—painful and liberating—she is here for a different reason. She is here to warn us that our fascination with the techno-utopianism of the Facebook screen will not lead to liberation but to cycles of anguish, of a body-politic in fissure.

Instead, her story—both personal and scholarly—asks us to enter a different body of light, not of the screen, but one of the soul accessible through the depths of meditation—of greater inner and outer connection and community.

We can panic or we can levitate. Where will we go? This book helps us find the way."

—Sohail Inayatullah, *Inaugural UNESCO Chair in Futures Studies, Tamkang University*

"What is the common morphology between the yogic-body and the smartphone-body? Between mediation and meditation lies the enigma of how we become humans by means of multiple anthropotechnologies that give shape to our yearnings of expansion toward limitlessness. Now the convergence of mind and screens are about to realize that everlasting but also precarious dream. In this process, the new social media technologies entail the promise of providing us with dazzling bodies of light.

A book impossible to leave alone until finished, *Does Social Media Have Limits?* proposes philosophy in the grand manner. The prose is mesmerizing and remarkably compelling. Impeccably crafted and with a highly focused reading, Mozzini-Alister sets the path for pursuing both the work of Vilém Flusser and Tantra Yoga's practical philosophy proposed by P. R. Sarkar. Rich in both intellectual and narrative rhythm, this book stands out in every respect.

A profoundly serious examination of the post-human condition in the age of social media, this research is as moving as it is insightful. Mozzini-Alister's incisive questioning is relevant for the fields of humanities, social media, and technology studies as well as it is indispensable for a new understanding of the meaning and value of meditation."

—Fabián Ludueña Romandini, *Professor of Philosophy, University of Buenos Aires, Argentina*

"Does Social Media Have Limits? reads like a scholarly piece with first-person reflections. Mozzini-Alister critically analyses the current obsession with social media technologies: acknowledging our daily struggles to constantly check our 'screens' in a desire to be omnipresent, to be superhuman. This book is easy to read and would appeal to scholars of social media, particularly those who study internet addictions."

—Pavica Sheldon, *Chair of Communication Arts at the University of Alabama in Huntsville*

"Camila Mozzini-Alister's book is an honest work about our societies' addiction to communication devices and social networks. The author focuses on the mediation between screens—or technical images after Vilém Flusser—and our minds, which is described from a media philosophical/archaeological perspective. However, Mozzini-Alister's work is not only a theoretical investigation but it also shows a spiritual and philosophical way out of being trapped in black-box-like apparatuses. She finds arguments for changing these habits in meditation and joyfully plays with the relationship between the terms 'mediation' and 'meditation'. Hereby a new aspect of Vilém Flusser's thought is also introduced to the reader: how Indian philosophical influences can be traced back in his media philosophy."

—Anita Jóri, *Research Associate at the Vilém Flusser Archive, Berlin University of the Arts, Germany*

"*Does Social Media Have Limits?* is a work of creative and confronting scholarship in which the profound parallels between mediation and meditation are explored. Consciousness and subjectivity will never be the same!"

—Marcus Bussey, *Professor of Futures Studies, University of the Sunshine Coast, Australia*

Camila Mozzini-Alister

Does Social Media Have Limits?

Bodies of Light & the Desire for Omnipresence

palgrave
macmillan

Camila Mozzini-Alister
University of the Sunshine Coast
Sippy Downs, QLD, Australia

ISBN 978-3-030-74119-8 ISBN 978-3-030-74120-4 (eBook)
https://doi.org/10.1007/978-3-030-74120-4

This Palgrave Macmillan imprint is published by the registered company Springer Nature Switzerland AG.
The registered company address is: Gewerbestrasse 11, 6330 Cham, Switzerland

To the innocence of all children
Especially the one who lives in our hearts
For Diana

Foreword

Camila Mozzini-Alister is an addict. I didn't know this when I met her in the Vilém Flusser Archive in Berlin in 2017, but she freely admits it in her new book "Does Social Media Have Limits? Bodies of Light & the Desire for Omnipresence." Mozzini-Alister writes, "although I was totally aware of the mechanisms of neoliberal capture at play in the use of social media and communications devices since I had studied them so thoroughly during my Research Master's Degree in Social and Institutional Psychology, I simply lost control."

And yet, we are all addicts nowadays. Addiction was embedded in earlier products: alcohol, sugar, and the excess salt, fat, and sugar in processed food. However, it is programmed into electronic devices in a particularly devious and cunning way. The technology at our fingertips has turned us into junkies, craving the eternal stream of information and entertainment available through social media, but also the next hit of dopamine released by our brains when we see a "like" on one of our posts. It's hard for the "real" world—the one that exists outside our devices—to compete with whatever is happening online. FOMO, the "Fear of Missing Out," flourishes within the digital sphere: if you put down that phone, you might miss a historic announcement, a revolution, or early news of a celebrity break-up. "Digital addiction" among children and teenagers has become a paramount concern.[1] In fact, many executives who work for tech companies are known to strictly monitor their own children's screen time. "We

[1] Dimitri A. Christakis, MD, MPH, "The Challenges of Defining and Studying 'Digital Addiction' in Children," *JAMA* 2019; 321(23): 2277–2278.

limit how much technology our kids use at home," Steve Jobs, a co-founder of Apple, Inc., once told a reporter for *The New York Times*.[2]

Where earlier, habit-forming products and substances came with warnings and, later, some sort of regulation about who could consume them, the algorithms in our social media and other platforms now read our patterns, stoking our desires and insecurities, coaxing us forward to read more, watch more, buy more. Experimental psychology was harnessed in the Stanford Behavior Design Lab that served as a pipeline to Silicon Valley to create patterns that undermined our freedom to choose. These, of course, have blossomed into a complex apparatus described as the "addiction economy," which relies on an asymmetrical exchange of information in which users surrender their private information for access to services and "data collectors, meanwhile, fiercely guard their own privacy, typically refusing to disclose what information they have, whom they will sell it to, and how they use it to manipulate our behavior."[3]

Shoshana Zubkoff calls this economy something even more sinister: a form of "surveillance capitalism" that "depends upon one-way-mirror operations engineered for our ignorance and wrapped in a fog of misdirection, euphemism and mendacity," in which the free, wild, and ungoverned space of the early Internet has been taken over by private surveillance capital.[4] What seemed like free services were provided by surveillance capitalists who regard us as free commodities: "We thought that we search Google, but now we understand that Google searches us," Zubkoff writes. "We assumed that we use social media to connect, but we learned that connection is how social media uses us."[5] The social inequality and "epistemic inequality"—knowledge, power, and access to technology—that the digital revolution proposed to erase have created even greater asymmetry—and in places like China, methods of state surveillance that reach deeply into private life.[6]

[2] Nick Bilton, "Steve Jobs Was a Low-Tech Parent," *The New York Times* (September 10, 2014): E2.

[3] Maya MacGuineas, "Capitalism's Addiction Problem," *The Atlantic* (April 2020): 11.

[4] Shoshana Zuboff, "You Are Now Remotely Controlled," *The New York Times,* January 24, 2020, https://www.nytimes.com/2020/01/24/opinion/sunday/surveillance-capitalism.html. See also Shoshana Zuboff, *The Age of Surveillance Capitalism: The Fight for a Human Future at the New Frontier of Power* (New York: Public Affairs, 2019).

[5] Shoshana Zuboff, "You Are Now Remotely Controlled."

[6] For instance, China's "social-credit" system based on data collected on individuals, such that citizens can lose the right to travel if they are caught jaywalking or playing music too loud. See Franklin Foer, "Beware the Digital Cure," *The Atlantic* (July/August 2020): 24–26.

The philosopher whose work Mozzini-Alister and I research, Vilém Flusser (1920–1991), predicted aspects of our current situation—or at least a world in which digital images would permeate our consciousness. Drawing from the phenomenology of Edmund Husserl and theories of information, communications, and visual art, Flusser foresaw a constantly shifting "universe of technical images" (technical images being anything produced by a technological apparatus, from chemical photographs to medical, scientific, or satellite imaging) and he imagined that we would communicate via screens—a practice that has reached critical mass during the global COVID-19 pandemic.

Flusser began his book *Into the Universe of Technical Images* (1985) with a "Warning" about the new society that would result from "synthetic" electronic images: "Current scientific, political, and artistic categories will hardly be recognizable there, and even our state of mind, our existential mood, will take on a new and strange coloration."[7] Earlier, in *Post-History* (1983)—the title reflected the change that would come about in an age in which linear writing (which produced causal logic and History) was eclipsed by images—Flusser wrote about "narcotic media" and wondered, "to what degree is inebriation found within the program of today's apparatus?"[8]

Flusser did not live long enough to experience the Internet, social media, or surveillance capitalism, but Mozzini-Alister picks up this task, addressing how what Flusser called our "existential mood" has been affected and how we can address the deleterious effects of digital electronics and social media on a personal level. Using a background grounded in academic research, but also a robust meditation practice, Mozzini-Alister proposes a solution. "My goal now," she writes in this book, "is to address the core of our current addiction to social media and help society to consciously shape our hegemonic technologies as a tool for self-knowledge in four different levels: personal, social, environmental and humanitarian, as well as spiritual." She does this by creatively imagining "mediation" and "meditation," examining the components of contemporary technical

[7] Vilém Flusser, *Into the Universe of Technical Images,* trans. Nancy Ann Roth (Minneapolis and London: University of Minnesota Press, 2011), 3. Originally published as *Ins Universum der technischen Bilder* (Göttingen, Germany: European Photography, 1985).

[8] Vilém Flusser, *Post-History*, trans. Rodrigo Maltez Novaes (Minneapolis: Univocal, 2013), 134. Originally published as *Pós-História – Vinte instantâneos e um modo de usar* (São Paulo: Duas Cidades, 1983).

images—screens, dots, and waves—and describing the "inner expansion" she has experienced through a disciplined practice of meditation. Mozzini-Alister considers what it might mean for humans to become "bodies of light": emancipated from "their mammalian condition," engaging with technology while acknowledging and diffusing its manipulative and addictive qualities. Where authors like Zubkoff follow the political trajectories of technology giants and how they have evaded regulation and eroded democracy, Mozzini-Alister offers a road-map for fixing our spiritual condition here and now in the Universe of Technical Images. "Ah! If we could unite the expansions of mediation and meditation, mind and screen, fun and discipline so that we could never again forget the mystery of who we are," she writes, "we would no longer have such a traumatized world." Here is an antidote not only to addiction, but also to the dystopias that have ascended with our technological "advancement."

Martha Schewendener
Instructor at New York University
Steinhardt School of Art and art critic
for The New York Times

Instructions for Reading

The Book

"Does Social Media Have Limits?" is a book with many possible layers of reading. It can be a manifesto, an invitation, a wake-up call, or just one more book in the library. Above all, it is an interdisciplinary investigation about how our desire for expansion—be it technological or be it spiritual—can impact our future as a species. To talk about such a complex subject, it was built a systematized structure. After the Introduction (Chap. 1), the following chapters (Chaps. 2 and 3) are, respectively, about social mediation and spiritual meditation or what was called "outer" and "inner" expansion. These two chapters were designed to work as complementary opposites, which means there is a symmetric game between them since one is the inverted mirror of the other. However, in the next chapter (Chap. 4), such logic changes. In it, the two parallel lines of mediation and meditation meet and melt into a "magic circle" when mediation becomes an addiction, and meditation, an obsession. In this moment, our desire for limitlessness is distorted into what I called "desire for omnipresence": a deep subconscious desire to get rid of our own animality in order to be as perfect as the images of our virtual profiles and spiritual gurus. What would be the consequences of such wish to be more-than-human? Are there limits to imagining our progress? Those are the kind of questions that the conclusion "Limits" (Chap. 5) will address.

The Reader

Dear Reader, you are the most valuable part of this book since it can only be completed by you and all the things that you bring with you. I will give you here some quick tips of how to approach this work, but feel free to create your own unique way of reading it. To begin, I would suggest, if you are an academic that seeks to understand the connections between mind and screen, mediation and meditation, outer and inner expansion, to read the full book, including the footnotes. However, if you have interest in such subjects but do not want to know the specificities of some technical academic aspects, I would recommend to skip the methodological section of the Introduction (Chap. 1) and have a quick look into the footnotes to see if they serve you or not. This book was written to be accessible to any reader and it is not restricted to specific audiences, so enjoy it the way you feel comfortable to do it.

If you already meditate, this is a great opportunity to deepen your knowledge and experience in such practice. However, if that is not your case, please don't feel uninvited: this book is not only for the initiated ones. It can also be for you, that barely heard of meditation, or has only a vague interest on the topic. Or even you that completely avoids it. Beyond the practice of meditation itself or even if you do or do not have any sort of metaphysical beliefs, the main role of meditation in this book is to enable us to go through the so-called *kos'as* or layers of the mind. There is no need to believe in the science of tantra if it doesn't make sense to you. Here, all I ask of you is to just give it a chance as a possible way of doing philosophy of mind and let it work within its function of being an inverted mirror of our current mediation practices.

For those who would like to add another layer of aesthetic experience to the reading of "Does Social Media Have Limits?," I would recommend listening to some Brazilian music with it. My favorite artist is Marisa Monte and the album "Verde, Anil, Amarelo, Cor-de-Rosa e Carvão" (1994). There is also Tribalistas, Caetano Veloso, Gilberto Gil, Tom Jobim, Elis Regina, Maria Bethania, Cidade Negra, Falamansa, Tim Maia, Cássia Eller, Nação Zumbi, Engenheiros do Hawaii, Chico César, Zé Ramalho, Lulu Santos, Skank, Paralamas do Sucesso, Vitor Ramil—or maybe listen to online radios such as Nova Brasil FM. Go with the flow. Let the Brazilian beat melt your heart and guide your soul. This will help you connect with the atmosphere of my world and the world of this book, cradled by the nostalgia ("saudade") from its Brazilian roots in the so far away Australia.

The First-Person Narrative

The last and final instruction is about the use of the first-person narrative. As a social scientist who devotedly seeks to understand the repetition patterns of human behavior, the decision to speak as an "I" involved a few different reasons. First, I wanted to give the reader the backstage, the conditions of possibility that have shaped me to develop this research. We all have our narratives, our unique ways of sculpting the facts of our lives into our own stories. No academic or scientist is an exception. Therefore, instead of neutrally describing the subject of this investigation, I wanted to allow the reader to understand my place of speech, the place where I came to see things the way they are presented here, in this moment. For me, this "knowledge about the knowledge" is a more than necessary exercise of academic honesty in order to engage not only with our own humanity, but also with the humanity of the other.

Second, academic texts can be distant, dry, unpersonal, and highly abstract. This is why I decided to use the "I" as a tool. I wanted to talk about desire for omnipresence in a desirable, seductive way. I wanted to experiment merging science and romance, object and subject, research and literature, content and form, complexity and simplicity. I wanted to provoke you, the reader, to come close to me. So close that you can actually feel the whispering of my words at your ears. My voice and words play a song. Try to build your own universe with it. Understand, but also feel it. Allow yourself to engage with each of the lines and sentences of this book as if they were your own "I" voice. Let go of the need to control and allow yourself to be guided into my/your/our story. Listen to this silent melody as if you had headphones on and everyone else around you was blindfolded. Here, you are safe. Trust me: it will be fun and, in the end, everything will make sense.

This is a dance.
You only need to give the first step.
Let's go?

Acknowledgments

Here goes my big "thank you" to:

The Coordination for the Improvement of Higher Education Personnel (CAPES), Research Support Foundation of the State of Rio de Janeiro (FAPERJ), Erasmus Practices (EP): for financially supporting this research.

Palgrave Macmillan, Philip Getz, Amy Invernizzy, Pavica Sheldon, Susan Westendorf, Dhanalakshmi Muralidharan, and Arun Prasath: for believing in this project and giving me all the support along the way.

Michel Foucault, Vilém Flusser, P. R. Sarkar, Fabian Ludueña Romandini, Walter Benjamin, Hans Ulrich Gumbrecht, and Jussi Parikka: for providing safe foundations to anchor the ideas of this book.

The University of the Sunshine Coast and Marcus Bussey: for opening doors for my work, allowing me to keep developing and growing my career in the other side of the world, and giving back to the Australian society my gratitude in being a part of this country.

All members of D.I.V.E. Theatre Collective, César Genaro, and Priscila Cunha: for creating such a necessary artistic space in the Sunshine Coast and welcoming my body and this book to become a part of the performance "Into the Light."

Erick Felinto de Oliveira and Josepa López Poquet: for, as PhD advisors, giving me all the trust, encouragement, and autonomy to develop my ideas during the PhD in Communication and Arts.

The State University of Rio de Janeiro and the Polytechnic University of Valencia and all the professors, colleagues, staff, and the research groups "Communication, Art and Sociotechnical Networks" (Comunicação, Arte e Redes Sociotécnicas, PPGCOM-UERJ), "Technological Cultures:

Medialities, Materialities, Temporalities" (Culturas Tecnológicas: Medialidades, Materialidades, Temporalidades, PPGCOM-UERJ), and "Mutant Screens: Interactive audiovisual projection with mapping techniques in unconventional spaces" (Pantallas Mutantes: La proyección audiovisual interactiva con técnicas de mapping en espacios no convencionales, (Laboluz-UPV): for welcoming me into such wonderful and alive environments.

Muniz Sodré, Rafael Diehl, Cíntia Sanmartin Fernandes, Denise Siqueira, Ana Taís Martins Portanova, Leticia Matheus, Cesar Baio, Charles Feitosa, Patricia Rebello, Emilio Martinez, Paula Sibilia, Fernando Antônio Resende, Fernando do Nascimento Gonçalves, Maribel Domenech, and Gustavo Bernardo Krause: for accepting to be a part of the many evaluation tribunals of this manuscript, helping me direct the paths to shape this research.

Inês Henningen and Rafael Diehl: for being my advisors at the Research Master in Social and Institutional Psychology with such commitment, knowledge, and care, allowing me to write the book "Impressions of a connected body: how advertising is inciting us to digital connection" which was the base of this research.

The Federal University of Rio Grande do Sul and the professors, colleagues, and staff of the Research Master in Social and Institutional Psychology: for providing me a strong Foucaultian base to anchor my thoughts.

The Federal University of Rio Grande do Sul, and all the professors, colleagues, and staff of the Biblioteconomy and Communication Faculty (Fabico), specially Ana Taís Martins Portanova Barros: for introducing me to the world of Journalism, academic research, and fancy words like "epistemology" in the company of thinkers like Edgar Morin, David Bohm, Boaventura de Sousa Santos, Illya Prigogine, Thomas Kuhn, and Paul Feyerabend.

Fábio Valenti Possamai and Nina Petek: for helping in the translation process of this book into English.

Thibéry Maillard and Thiago Krening: for beautifully digitalizing my drawings.

Alexandre Melo, Elli Adams, Götz Mundle, Flávia Gonzalez, Eliana Zákis, Peter Burge, Deepti Rampal, Poobalan Naidoo, Carmen Comin, Lakshimi Subramanian, Maria Inês Garbino Rodrigues, Yildiz Sethi, Jorge Lopez, Dean O'Brien, José Ricardo, Marc Daniels, Connie Page, Letícia Figueiredo, Patrícia Flores, and Jagath Bandara: for helping me to keep my physical and mental health amidst the difficulties of a double PhD and the adaptation process in Spain, Germany, and Australia.

Vilém Flluser Archive (Universität der Künste Berlin) and the Flusserian Family, Anita Jóri, Alexander Schindler, Rafael Alonso, Martha Schewendener, and Lucas Maia: for existing and occupying the warmest place of my heart.

Samira Franz Boligian, Marilene Batista, Angela Bagattini, Charlotte Bernaulte, Juan Antonio Medina, Helen Santos, Fernanda Bassani, Aline Schmit, José Henrique, Paula Bianchi, Daniel Silveira, Júnior Lopes, Joseane Flor, Rosilda Santos, Teresinha Jorvino, Isabella Joyce, Grécia Falcão, Helena Klang, Filipe Feijó, Fabiola Lorenço, Alessandra Maia, Lucas Neiva, Amanda Torres, Paulo Souto Maior, Olga Ballester, Nitya Dambiec, Kalyan Pérez, Raphael Meissner, Larissa Krause, Riccarda Freund, Maximiliam Rest, Ana Lucia Pareja Barroso, Carla Venturini, Francisco Santos, Celina Martins, Fabiana Daniel Felicio, Mariana Conte, Tatiane Cláudio, Hellen Duarte, Carla Camila Moreira, Débora Jacques, Flávia Azambuja, Joana Amaral, Mariam Jamil, Ivan Galaxto, Volkan Knopfler, Maria Cristina Lodoño, Davy Guedes, Adil Es, Lucia Antonini, Luíze Bueno, Juliana Winckler, Ana Negreiros, Nina Hilmann, Alyona Ved, Valeria Flores, Mariana Faro, Paul Cercueil, Etiane Araldi, Elis Mori, Susan Andrews, Robertha Oliveira, Tiago Pires, Renato Dombrowski, Liana Bohn, Renata Grunnenvaldt, Angelo Reruffo Rodrigues, Ana Paula Cristofolini Rocha, Tatiane Dotta, Sasha Newman, Camila Ribeiro, Regina De Miguel, Sandra Burnett, Chantelle Louise, Fernanda Gomes, Elmir Matheus, Jéssica Gogan, Raphael Giammattey, Clarisse Fonseca, Clarice Greco, Lilian Hack, Iliriana Rodrigues, Mariana Rosa, Dalva Bavaresco, Dasha Lavrenikov, Ludi Bobbe, Luiz Guilherme Vergara, Vanessa Costa, Manoela Resek, Nathalie Felicio, Gilson Do Vale, Flor Do Vale Mohajer, Luciele Vicente, Anne Shannon, Steven Michael, and all the friendships created all over the world, specially in Porto Alegre, Córdoba, São Paulo, Rio de Janeiro, João Pessoa, Valencia, Berlin, Melbourne, Brisbane, and the Sunshine Coast: for the unconditional trust, truthfulness, and support.

Rosângela Vilar and Marcos Espindola: for being my favorite "crazies" and for everything you have done for me in order to have my PhD thesis defended.

Facebook Mediation and Tantric Meditation: for the inner and outer expansion of my consciousness and for allowing me to have so many beautiful encounters in life.

Kunz, Mozzini, Marcon, and Milani Family: I carry the flame of your unconditional love in every cell of my body. Mom, Dad, Talissa, Henrique, Gradma Teozene, Grandma Jesus (in memoriam) Grandpa Davino (in memoriam), Grandpa Henrique (in memoriam), Paulo, Davi, Évora, Helena and Pedro, and all my Ancestors, Uncles, Aunties, and Cousins: for giving me life, encouragement, and support all along.

O'Callaghan, Fitzgerald, Alister, Bussey, and Jeays Families: for sculpting the love of my life and welcoming me in your lives and hearts.

Baba, Jesus, and Mary: for being the eternal mystery guiding my heart.

Nature: for being the physical expression of God/ess' purest love and compassion that has always renewed my energy to never give up.

Brazilian Music: for giving me the strength to overcome any obstacles.

Ajit Nicholas Mozzini-Alister: for being by my side all along.

Contents

About the Author

Camila Mozzini-Alister works as a Social Media Educator and an Adjunct Research Fellow at the University of the Sunshine Coast (USC), Australia. Her background is a PhD in Communication from the State University of Rio de Janeiro (UERJ-Brazil) and a PhD in Arts: Production and Investigation from the Polytechnic University of Valencia (UPV-Spain), with a Research Master's in Social and Institutional Psychology from the Federal University of Rio Grande do Sul (UFRGS-Brazil) and a Bachelor's Degree (Honours) in Communication—Journalism from the same institution. Her research interests rely on the interfaces between body, philosophy of mind, social media, digital mediation, tantric meditation, desire for omnipresence, affection, migration, education, as well as her work as a performing artist. Throughout her career, she worked as an Art Educator at both Iberê Camargo and the Mercosul Biennial Foundations, as well as a Social Educator at the Mathias Velho House of Youth, where she taught Citizen Journalism to the vulnerable youth of the community. She co-edited the books "Body Studies: Encounters with Art and Education" (Estudos do Corpo: Encontros com Arte e Educação) in partnership with Wagner Ferraz, and "Phenomenologies of Grace: The Body, Embodiment, and Transformative Futures," with Marcus Bussey. In 2019, she published her first monograph titled "Impressions of a connected body: how advertising is inciting us to digital connection" (Impressões de um corpo conectado: como a publicidade está nos incitando à conexão digital). Camila takes writing as a field of poetic experimentation and the body as our first and ultimate creative substratum.

List of Figures

CHAPTER 1

Introduction

1.1 Affections as a Research

Migrating is a creative yet painful situation.

A whole literature deals with the relationship between creativity and suffering. Those who leave their homeland (by necessity or decision, and both are difficult to separate) suffer. Because a thousand threads bind them to the motherland, and when they are amputated, it is as if a surgical intervention has taken place. When I was expelled from Prague (or when I made the courageous decision to flee), I experienced the collapse of the universe. What happened is that I confused my inner self with the space outside. I suffered the pains of amputated wires. But later, in the London of the first years of the war, and with the premonition of the horror of the camps, I began to realize that such pains were not related to surgical operation, but to child birth. I realized that those cut wires had nourished me, and that I was being designed for freedom. I was taken over by the vertigo of freedom, which is manifested by the reversal of the question "free from what?" into "free to do what?".

And so we migrants all are: beings taken by vertigo.

Vilém Flusser

The presence of the image has always intrigued me. I grew up in the decade that saw the complete transition from the analogic to the digital. My records turned into cassettes, cassettes became CDs, CDs were converted into disks, disks into flash drives, flash drives into Wetransfer and Google Drive. I remember the day when my parents bought our first

C. Mozzini-Alister, *Does Social Media Have Limits?*,
https://doi.org/10.1007/978-3-030-74120-4_1

computer, an immense contraption that ran on Windows 95. I was only eight years old and that seemed like a new world to me. While I was playing with coloring games, my two older siblings would access MSN and clutter our telephone bill with chat services fees. I also remember my first cell phone. It was a Samsung Slim. I was one of the last persons in my classroom to enjoy this technology, which was then seen as a symbol of status among my classmates. "The last shall be first," as the saying goes. Well, at least in this case this held true: my phone was one of the most modern in the classroom and I finally felt like I belonged to the "technologically included" ones. It was always hard for me to keep up with the speed of these transformations. I felt as if everything was changing too fast and all that new information would be, in the end, useless since it would be discarded again and again in the endless cycles of planned obsolescence. I was more interested in wandering: to philosophize, to digress, to dance, and to draw. I am neither the best nor the most updated person to talk about technology. I do not know how to code and I barely know how to use photo editors. My computer is a Dell with such a small memory capacity that, right out of the gate, my backwardness in the higher echelons of media art investigation across the world is quite clear. Driven by "Ferraris" from Apple, my "old Volkswagen Beetle" and I take such "inefficiency" as a starting point for the study of our relationship with the current image and communication devices: a starting point less functional and naturalized, more affective and existential.

After I began my Ph.D., I adhered to the use of smartphones with touchscreen technology. In the last four years (2014–2018), I saw my habits and my daily life been drastically altered by the presence of this new being, this new intruder that, unbeknownst to me, has conquered so much space in my life. I became, little by little, addicted to gliding my fingertips on photosensitive screens. Even when I had not received any messages, I was still constantly checking and getting lost in social media. Basically, this was the brilliant focus of my "smartphone" usage. Every fifteen minutes, it was as if an inner alarm would set off and I had to immediately access those communication platforms: what was going on in my Facebook? Was there somebody calling me on WhatsApp? I was conditioned: there was an incontrollable sense of urgency… Even now, at this very moment, such mental alarm clock is ringing while I am writing these words. Only I know the effort I must make to be able to resist the excess of accesses to which I feel invited to partake in. And it is not like I didn't had anything to do: I had, and a lot! However, I saw myself slowly automatizing a new necessity of presence. A strange presence: although I was totally aware of the mechanisms of neoliberal capture at play in the use of social media and communications devices since I had studied them so thoroughly during my Research Master's Degree in Social and Institutional Psychology, I simply lost control. Since then, my days became filled with fierce battles against symptoms such as: difficulty to concentrate on daily tasks due to constant glances at my phone, reluctance to interrupt the cyclic gliding of fingers on photosensitive screens, and anxiety to talk to people that were not close to my body: what was making me so entangled in such practices?

Perhaps my gypsy past was a catalyst in making this situation a reality. I never knew what was like to lay deep roots in a stable ground or house: I'm a descendant of Italian and German grandparents who moved to Brazil and mingled with the blood of Native and Black people there and whose grandsons and granddaughters—my parents—also migrated from the South to the Brazilian Northeast when I was only four years old. This is a good moment to have a map of Brazil to realize how far this is. From my birth until I was four years old, I lived in the city of Tapejara, in the countryside of Rio Grande do Sul. From four to ten years old, in Campina Grande, in the state of Paraíba. From ten to eighteen years of age, in Paraíba's capital of João Pessoa. From eighteen to twenty-six years of age, I came back to Porto Alegre to study. In the meantime, I lived for six months in Córdoba, Argentina. From twenty-six to twenty-nine years of age, I moved to Rio de Janeiro in order to pursue my Ph.D. studies. When I was twenty-nine years old, I went to Valencia, in Spain, with the dream of doing a co-tutelle with the Polytechnic University of Valencia. I lived there for nine months and then I stayed in Berlin for three months, where I worked as a Research Fellow at the Vilém Flusser Archive. Those were twelve intense months. Even for those used to the length of distances, to go to another continent, this time the "Old" one, and staying there for twelve months without feeling the warmth of a hug from the ones you love, hurts. I felt the pain of this old partner called "loneliness" becoming sharp. Sharp not only because I was distant from my roots, from my relatives, from my friends, and from the sweetness of the Portuguese language, but also due to the most unexpected encounter of souls that would implant in my life the longest of all distances: Australia. What was already a long one year away from Brazil, suddenly became two: I married an Australian and we decided to make the "brand new" continent our home, from where I now type at a café in the small district of

Frankston, the keys of my obsolete Dell in order to give shape to these words.

There is a saying in the academic world that goes "research problem, researcher problem." Everybody has their own existential hole. Mine is called "distance." Educated by the choices that I took as mine, unwillingly I became specialized in the theme of distances which, by consequence, brought the issue of mediations to the center of my life. The experience of distance subjectively constitutes me: to give up the physical touch of my loved ones in favor of pursuing my studies, or to be next to other loved beings, have always been a constant to me. And it is at this point that the presence of image enters the scene. Since a very young age, social media was the only way of bringing closer the ones who were no longer by my side. By inhabiting a body mediated by other forms of "flesh," I learned to grasp the other through the materiality of image. Such discrepancy from what is taken as "real" has enmeshed in me lenses that allowed me to see the present through the "eyes of the affection": I see the emergence of imagetic forms of presence as a central axis in the current modes of living, and not something that is merely "a problem of mine." I got tired of witnessing scene after scene of moments of pause being filled with the gliding of fingers on screens tuned to Facebook, Instagram, YouTube, TikTok, Snapchat, or WhatsApp, of people going out and about with their phones glued to their hands, of people on mass transit traveling also via the speed of Wi-Fi signals and data networks, of bar tables populated by drinks and noisy devices that would rudely insert themselves into the conversation, of mothers that, while taking their kids for a stroll, would talk to the image of people on screens. Why is it no longer enough being in the ordinariness of the offline here and now? What is it that escapes in this process of exchanging our personal data for the endless desire to communicate?

It was only after I incessantly ask myself why in blazes I am still on Facebook, even though I was fully aware of the marketing policies and subjective training, that a new angle of this problem took shape: the desire for omnipresence. I wonder if that is the reason why more and more people ignore or take for granted the obscure uses of their digital traces in exchange for the production of forms of presence that are not limited to the scope and reach of the human body. A desire for simultaneously inhabiting distinct space-times, for concomitantly being here and there, in the space of the physical body and in the space of the mediated body. A desire for being more than just a body. A desire for overcoming the very own human condition. A desire for extending the perimeter of arms, legs, and vision. A desire for being superhuman. A desire for becoming divine? A deification that, since 2014, when I began my Ph.D. in Communication at State University of Rio de Janeiro (UERJ), has allied the necessity of mediation to the disciplined practice of meditation.

Meditation was something that entered my life in an unexpected way. Despite my strong Christian roots, I found in the Yoga universe something that fascinated and attracted me. As hard as I tried to resist it, it was inevitable. I was initiated into Tantra Yoga, an offshoot of Shivaism whose ultimate goal is the expansion of all human potentialities. Quite rapidly, my daily life was filled with ásanas, mantras, the study of *cakras*, vegetarian food, breathing techniques and, twice a day, at least twenty minutes of meditation. I could feel the effects on my body. I felt the acceleration of my life to levels I have never reached before. I wanted to achieve the much-promised liberation of the mind. I aspired to become an "angel," to be a body of light and travel in the same way the Wi-Fi waves of a touchscreen phone would transport my body to another space and time. Mediation and meditation became, then, the two main activities that would introduce my body to the daily habits cultivated during the last four years of my doctorate studies. Media(t)ation. Two activities that decisively shaped and sculpted my perception of reality.

Nevertheless, it took me more than three years to not only realize, but also to accept the way that these two practices were intertwined with my path as a researcher. Unconsciously imbued with certain academic preconceptions, I avoided all forms and every vestige of what may be classified in the category of "spirituality." I was afraid that my text would be judged as "mystical" and, therefore, "groundless." After countless and countless crises, I could only be true to myself and put into practice such honesty throughout the process of writing. I had to fight off the desire to fill all the empty spaces in my arguments: the many authors and the thousands of pages and the books and articles I read during my Ph.D. did not articulate a hypothesis to be corroborated. There was no way to prove something that belongs to the order of the fragile, the vulnerable, the incomplete, since it comes from experience. Experience of the extrapolation of every possible limit of medi(t)ation on my own body that, in its blind search for expansion, found in these words the challenge to create a middle ground.

1.2 The Research of Affections

There is, in human beings, a thirst for the limitlessness.
P. R. Sarkar

There we were, Ajit (my Australian partner) and I, visiting the Vatican on July 2017. After a long walk, we reached our final destination: the Sistine Chapel. Although full of tourists, there was an air of silence around us. Gravity presented itself in its grave tone: Michelangelo frescoes on the ceiling silenced the human noise by speaking the mute language of the sacred. And it was without uttering a single word that Ajit made an unexpected connection: he pointed up, showing me the masterpiece "The Creation of Adam" and then touched the screen on his phone. When faced with the resemblance of his touching gesture, I was overcome by an astonishment: would we be mediating our relationship to the divine through screens? If so, what is the relationship between the mediation of screens and the practices of meditation? And what do those gestures have to do with the way social media is affecting us? The Yoga's maxim reverberated in my thoughts. It does not matter how diverse the many spiritual offshoots emerging from India may be, in all of them there is a common premise: "you are not your body." How they reached such conclusion, it is not up to me to explain in detail here, but it is a fact that, for the yogi wisdom, the physical body is just one of the mind's many layers of energy, and by the way, the outermost and least subtle. A logic that is being strangely re-enacted on the current smartphones with touchscreen technology: instead of buttons and keys protected with an isolating material, nowadays we transport the body via radiofrequency waves that travel by the direct exchanges of electric current between photosensitive screens and fingertips. There is something singular in this: for the first time, the humanity of our techniques electromagnetically inserts the body into other layers of presence no longer restricted to the field of the visible. A search. A search to reach something that is beyond the corporeal substratum, something that ruptures the tridimensional conventions of space and time, something that allows us to be simultaneously in every "here" and every "now." Something that makes us, at last: omnipresent. What would come out from the meeting between mediation and meditation?

Before starting our journey toward the investigation of this question, let us be specific with the meaning of a few concepts that will be recurrent here: (a) desire for limitlessness; (b) desire for omnipresence, (c) image, (d) meditation, and (e) mediation. Starting with the term that has become a commonplace in any advertisement slogan that incites us to be digitally connected: by (a) desire for limitlessness we will understand the innate spiritual desire to expand our consciousness toward the infinite (Ánandamúrti, 1993). Driven by such desire, our current (b) desire for omnipresence is comprehended as the desire "to reach a dimension in which the tridimensional experiences of space and time are no longer applied." Said dimension is usually related to the field of the sacred, the divine: God. Pertaining to (c) image, it will be understood as "a means of expansion of space and time." Finally, we have the two fields employed as the basis for our study of the current desire for omnipresence: (d) meditation: "direct process of inner expansion towards the limitlessness through the subtilization of the mind." To that end, techniques that subtilize the body, such as vegetarianism, *ásanas*, and mantras, seek to control the lower *cakras* in order to enable us to tune with wave lengths invisible to the human eye; and (d) "mediation": "indirect process of outer expansion towards the limitlessness through the subtilization of the body as an image of a virtual profile." To that end, techniques of interaction in social networks, such as newsfeed scrolling and the sharing of thoughts and emotions prompt us to create forms of energetic symbiosis between body and device.

We will study these two modes of expansion of the body's presential limits, meditation and mediation, using the concept of anthropotechnics, that is, "techniques through which human communities and the individuals who compose them act upon their own animal nature with the goal of guiding, expanding, modifying or domesticating their biological substratum" (Ludueña Romandini, 2012, p. 9). Nevertheless, since "every anthropotechnology implies a zoopolitical substratum that lies at its center" (Ludueña Romandini, 2012, p. 10), we will enter in the delicate realm of the *zoé*, that is, of the spectrality. Spectrum, for Ludueña Romandini (2012), relates to two different meanings: in a broad sense, it pertains to the "incorporeal creatures like, for instance, the angels" (idem) and God, with a capital "G", that "is expressed through the Spirit" (idem) and carries a common ontological ground between Spirit and Spectrum. In a strict sense, it relates to the "beings that survive (even though in the form of a postulate) their own death, or that establish a point of indistinction between life and death" (Ludueña Romandini, 2012, p. 13). Under this light, "the spectrum can be completely immaterial or acquire different 'consistencies'" (idem).

The motive for including this perspective is due to the fact that without the study of the "spectral dimension, it is impossible to have a cabal comprehension, not only of the contemporary political drift in general, and of the zoopolitics in particular, but also of the new possibilities of ontology" (Ludueña Romandini, 2012, p. 12). Different ontologies that, avoiding questions like "what is this?", move toward questions such as "how is this produced?" and "how is this possible?". Questions that increasingly operate a self-criticism of our own selves by a problematization of the ways through which we build ourselves as "humans," and no longer "animals." In this sense, we will dare to dwell on the study of the contemporary "spectral climate" (Flusser, 2008) from the perspective of Tantra Yoga's subtle knowledge. This choice was made because the very goal of Tantra is intimately related to the hard core of the concept of anthropotechnic as an expansion and domestication method of the biological substratum: "the spiritual practice which liberates the aspirant from the dullness or animality of the static force and expands the aspirant's [spiritual] self is Tantra"[1] (Ánandamúrti, 1993, p. 167).

Thus, taking into account the immense supply of contemporary anthropotechnics, such as the "new biotechnology, A.I., nanotechnology, molecular biology, physics and computer and cognitive sciences"[2] (p. 224), we will use the binomial mediation/meditation as elements to dissect the current modes of deconstruction and therefore, expansion, of the body's physical form. Body, in the practices of medi(t)ation, reconfigures its anatomy based on the electromagnetic spectrum manipulation techniques or, in yogic language, "*prana*." Whether being in the search for the mind's enlightenment through the subtilization of the body, or in the body's electrocution through the conditioning of the mind to the constant touching of screens tuned into virtual profiles, we will navigate these two apparently distant universes. Our goal is to investigate the modes of affection, that is, the effects of tangibility arising from bodies that are no longer restricted to one taken as "the solid, the tangible, what is sensitive and,

[1] The complete passage goes: "The significance of the term tantra is 'liberation from bondage [the bondage of dullness, or staticity]'. The letter ta is the seed [sound] of dullness. And the root verb trae suffixed by ḍa becomes tra, which means 'that which liberates'—so the spiritual practice which liberates the aspirant from the dullness or animality of the static force and expands the aspirant's [spiritual] self is Tantra sádhaná. So there cannot be any spiritual practice without Tantra. Tantra is sádhaná."

[2] It is worth mentioning that Ludueña Romandini (2012) does not consider anthropotechnology to be an "inexorable process of fabrication of what is human that takes us from 'the classical humanism' towards a new anthropotechnological era of biotechnological eugenics" (p. 10). Contrary to a kind of deontological premise with a predetermined target, the modes of anthropotechnological substantiation are related to a "*contingent form* adopted by the very technologies of power applied on the human animal" (idem).

Fig. 1.1 The screen mediating the divine touch, by the author

mainly, to that which is bathed in light and is, therefore, quite visible" (Dagognet, 2012, p. 2). Affections that, deviating from their usual meaning related to the feeling of care toward something or somebody, allow us to talk about the potential of the material impact produced by relationships—as Deleuze would contemplate (1978) when talking about Espinosa—between bodies, whether they are made of flesh or light (Fig. 1.1).

1.3 Method for the Research of Affections

My body is implicated in everything.
Maybe my liver less than my eye, but I am not even sure about that.
Every analysis of the world will reveal, in the end, my body. [...]
It is paramount to create a phenomenology of the body,
before creating no matter what kind of Christology.
Vilém Flusser

By proposing as our research problem the question of "how the desire for omnipresence produces modes of expansion of the biological substratum?", we will anchor this investigation on the premise that there is, in the current practices of mediation, something that goes beyond the hypothesis of a market subjectivity cooptation. Hence, this phenomenon will be undertaken in its productive dimension which, instead of simply preventing, forbidding, surrendering, obliging, or determining, also constitutes an action upon action that "operates on the field of possibilities in which the behaviour of active subjects is able to inscribe itself" (Foucault, 2010b,

p. 288). Therefore, we will seek in the act of uninterruptedly touching screens the search for a common desire to the practice of meditation: the desire for expansion toward the limitlessness, or enlightenment. Thus, taking into account the general goal of "investigating the affections at play in the biological substratum's modes of expansion," three specific goals will be developed: (1) to systematize the process of outer expansion found in the practices of mediation based on a material archaeology of the screen; (2) to systematize the process of inner expansion in the practices of meditation based on a material archaeology of the mind; (3) to operate the intersection between both practices of expansion so that we can systematize the different degrees of affective programming that actualize the anthropotechnological project of transforming the body into a body of light. Thereunto, we will organize our journey through the interlocution of three chapters: (1) mediation, (2) meditation, and (3) medi(t)ation.

Our argumentative background shares the common soil permeated by the maxim of the Ladder of Abstraction (Flusser, 2008). In it, Flusser states that for us to become aware of the concreteness of the world, we would first take distance from it in order to grasp it better via intermediates of things called "technical images" (Flusser, 2008). As manifestations of the highest degree of abstraction in relation to the concrete world, technical images would be, in a nutshell, images that do not need the direct action of the human hand due to the mediation of devices that work just like black boxes: hermetic, automatized, and programmed. Subjectively secularized and, therefore, far away from the possibility of grasping the experience of the divine, the limitlessness, in the banality of the molecular level of matter, today we desperate this search by the use of abstractions that are materialized into the form of two contemporary mediums of the biological substratum expansion: mind and screen. Immersed in the emptiness of the cosmic and topological spaces, we will propose a material archaeology of both mind and screen based on the proposition of a media archaeology (Parikka, 2012). As a method for investigating the "deep layers of the body" (Parikka, 2012, p. 30) inscribed in the non-visual field of algorithms, semiconductors, and in technologies such as the Wi-Fi, Bluetooth, UMTS, EDGE, HSPA, GSM, and GPS—and why not, also *cakras*?—we will apply it to the search for what I call "desire for omnipresence." A desire that can only be touched and manipulated through a methodological approach "that possesses a fierce transdisciplinary nature in which the alternative histories of the culture of mediums are pursued in some diffuse part on the border between art/science/technology" (p. 14).

Such method is inspired by "the studies in archaeologies of power and knowledge of Michel Foucault (1926-1984), the early excavations into the rubbles of modernity by Walter Benjamin (1892-1940), the New Film History in the 1980s" (Parikka, 2012, p. 3), as well as various studies which, since the 1990s, have sought to understand digital and software cultures with the help of the past, a layered "unconscious" of technical media culture. For that, a media archaeology links the notion of archive (Foucault, 2010c) to the notion of device, and look at our technologies as more than just technologies: they incorporate complex devices and tools for thought while immersed in a social platform of memory (Parikka, 2012) capable of producing affections. "Affections" that are entering more and more into the cognitive field, emerging as strategic tools to reflect about how the human body and mediation systems are creating processes of electronic symbiosis (Grusin, 2010). In this sense, to talk about the agency between mind and screen implies stirring a controversial and intense debate: how do we learn, apprehend, and engage with the environment that impels us, and is at the same time optically processed as images made of rays (Flusser & Bec, 2011)? Are we mentally related to a world that is already completed or do we create it based on the relationship between world and sensorial processes? Ultimately, where can mind and screen be found among that which we internally imagine and technically exteriorize as images? Bringing these questions to the center of the cognitive quarrel along the lines of the "Screen Consciousness Studies," Pepperell (2006) situates the body-mind dialectics by employing the classic dispute between internalistic and externalistic positions (Pepperell, 2006). Nevertheless, recognizing not only the impossibility, but also the unproductivity of choosing a side while overlooking the other, the author comes up with a third avenue: to open up the fissures of the paradox, the relationship between the complementary opposites, and circularity. A path that allow us to explore the relationship between screen, mind, and world using "transdisciplinary modes of inquiry, seeking insights from all branches of the humanities, sciences and arts, and perhaps more esoteric forms of knowledge that still need to be academically sanctioned" (Pepperell, 2006, p. 194).

And why not engender an analysis of this phenomenon based on the experimental science of yoga (Flusser, 2010) that has been utilizing, for thousands of years, the body to empirically dissect the cognitive mechanisms of the mind? There are many texts in which Flusser links together mind and technical mediums using the yogi tool set: works like "O último juízo: Gerações" (2017—unpublished in English), "History of the Devil" (2010), "On Doubt" (2011), and even "The universe of the technical images" (2008) present a few clues as to how much of the dialectics of his thoughts borrowed aspects from the pragmatic philosophy of meditation. Thus, by honoring the biographical and philosophical journey of Vilém Flusser, we will create a media archaeology of mind and screen by articulating three common axes in the material process of the abstraction from the concrete: shocks, points, and waves. Not coincidentally, these three discursive regularities[3] emerge in both practices of mediation and meditation: they crystalize points of convergence, recurrence, and correlation between the social production of reality and phenomena that are, at first, trivial and spread out across space and time. However, just like Flusser and Bec (2011) explored the abysses that separated and brought together humans and the rare species of the cephalopoda genus *Vampyroteuthis infernalis* from regularities that were symmetrically organized, we will employ these variables through the formal logic of an inverted mirror (Bernardo, 2011). That because, if by mediating we feel the shock between fingertips and screen that allows us to inhabit the imagetic universe of points such as bits and pixels, which are transmitted through waves and antennas, by meditating we do the symmetric opposite. First, we use *ásanas* and diets to subtilize the body's vibrational frequency, then we sit down to meditate in order to balance our *cakra* points and, at last, we receive the shocks of *samádhi*. In this mimetic game, *Vampyroteuthis infernalis*[4] can also be an inspiration for an experimental and discursive journey in which personal narrative and imagination are employed as a fuel for an epistemological search: if Flusser creates the "philosophical fiction"

[3] Far from being understood as something that is being discussed or as a "phenomenon of expression" (Foucault, 2010b, p. 61), the discourse implies here an open territory of practices in which one searches for "a field of regularity for several positions of subjectivity" (idem). Avoiding the concept of discourse as mere words, phrases, or propositions, it "presents intrinsic regularities to itself, through which it is possible to define its own conceptual network" (Fischer, 2001, p. 200). Hence, beyond the analysis of lexical content, the semantic structure, and the linguistics of signs related to words or things, the discourse for Foucault is "more than just utilizing such signs to designate things. It is this *more than* what makes them irreducible to language and to speech. And it is this 'more than' that one must make visible and that one must describe" (Foucault, 2010a, p. 55).

[4] In this work, Flusser and Bec (2011) perform an imaginative exercise that seeks to, skirting the dichotomies that separate and bring together humans and the rare species of the cephalopoda genus *Vampyroteuthis infernalis*, mirror the human existence from the perspective of this animal that lives where we die, and that dies where it is allowed for us to live.

(Felinto & Santaella, 2012) as a singular narrative genre, this happens not only because he appreciates the poeticism of allegories. Situated within the philosophical and scientific tradition of mental experiments, Flusser, like Foucault (1994)[5], brings to the surface a peculiar methodological approach: the alliance between scientific rigor and imaginary exploratory powers.

Thus, aiming to make visible the programming modes that impact the human body's most animal affections, we will choreograph the project of a *vampyroteuthic* morphology between the yogic-body and the smartphone-body utilizing a structure of mutual support: from the meditation practices, we will incorporate a processual system to the non-systemic practice of mediation; from the mediation practices, we will transplant the oscillation movement of the technical image into the systemic nature of the meditation practices. As pillars of mutual support, there is no way to understand one chapter without the other. Therefore, by "flusserianizing" the meditation of the mind and structuring the mediation of the screen, we will move on to the final chapter in which mind and screen finally materialize the desire for omnipresence while diluting themselves toward the conquest of the limitlessness: the perfect and omnipresent model of the resurrected body of Christ (Ludueña Romandini, 2012). As "the outcome of the first and truly technical and biological experiment of the divine zoopolitics" (Ludueña Romandini, 2012, p. 187), the Messiah consecrates in his resurrection the power of immortality: "man and God, born (*natum*) and not born (*non natum*), carnal (*carneum*) and spiritual (*spiritalem*), weak and strong, mortal (*morientem*) and living (*viventem*)" (Tertuliano apud Romandini 2012, p. 130). "After all, what is a resurrected body other than a living dead?" (Ludueña Romandini, 2012, p. 127). And against the distention of the zoopolitical life/death binomial in favor of the creation of surviving modes that modify the very matter of what is understood as a body, we will investigate, in the intersection

[5] In 1977, interviewed by Lucette Finas for *La Quinzaine Littéraire*, Foucault gave a categorical response when questioned about the fictional character related to his texts: "Pertaining to the question of fiction, it is to me a very important issue; I clearly realize that I have never written anything but fiction. I do not want to say with this that my writings were outside of the truth. It seems to me that it is possible to use fiction within the confines of the truth, to induce effects of truth with a fictional discourse, and to make it in such a way that the discourse of truth will create, will produce, something that does not exist yet, thus 'fictionalizing it.' One 'fictionalizes' history based on a political truth that makes it true, one 'fictionalizes' a politics that does not exist yet based on a historical truth."

between mediation and meditation, the seeds for the ascension of contemporary bodies of light.

To that end, we will study this current manifestation of the desire for limitlessness based on what authors such as Hayles (2006) and Thrift (2004) described as "technological unconsciousness." Technological unconsciousness is nothing more than "the everyday habits initiated, regulated, and disciplined by multiple strata of technological devices and inventions, ranging from an artifact as ordinary as a wristwatch to the extensive and pervasive effects of World Wide Web" (Hayles, 2006, p. 138). Entrenched in practices such as "scrolling, interacting with computerized voices, talking and text messaging on phones, and searching the web to find whatever information is needed at the moment" (Hayles, 2006, p. 140), the desire to exist beyond the limits of our own body becomes more and more intertwined with the technical unconscious of smart devices. Enmeshed within this flow of data, human behavior is increasingly integrated with a wide variety of other cognitive activities that "fall below the threshold of conscious awareness" (idem). Investigations such as those performed by Ariely (2010) and Mlodinow (2008, 2012) have already brought to light our irrational condition by finding that around 95% of our choices are unconscious. Thus, it is no accident that Grusin (2010) states that several gestures incorporated into our mediation practices cannot be known "not because they have been repressed or sublimated but because they are concealed or invisible or unrecognized in everyday practices that we participate in and take for granted" (Grusin, 2010, p. 71). Therefore, how can we bring to the surface that which cannot be found on the tip of the iceberg and, at the same time, is manifested through increasingly automatized practices?

To deal with this that wanders around the basement of the unconsciousness and, therefore, the occult, unknown, and unrecognizable, we will bring together the desire for omnipresence found in the practices of mediation and meditation. And since every archaeological endeavor requires a delicate journey, we will advance to the material excavation of the relationships between mind and screen utilizing the field of experience. Experience, from the Latin "*experientia*," composed by the prefix "*ex*" (to separate from the interior), the root word "*peri*" (to try, to attempt) and the suffix "*entia*" (the quality of a being), and thus related to the quality of proving or experimenting based on certain practices or elements. In this sense, "experience, as its etymology suggests, is a continuous passage through threats and dangers" (Shusterman, 1998, p. 267) that brings to the surface the articulation of three vital axes: "first, forms of a possible knowledge; second, the normative matrices of comportment for individuals; and finally, modes of virtual existence for possible subjects" (Foucault, 2011a, p. 4). It is no accident that Foucault affirmed that "knowledge is not made for understanding; it is made for cutting" (Foucault, 2011b, p. 28). It is by placing "experience as the generating matrix" (p. 6) of something that Foucault, during his life and work called knowledge, power, and the ethics of the self, we will methodologically advance the investigation of the sensorial and material impact of the medi(t)ated affections through the experience of the researcher-body. Body that, here, is the guinea pig of its own experiment and, facing the anguish of living a life mediated by distances, has found on social media platforms more than a mere entertainment: on it one can also find regularities, unconscious affective patterns capable of crystalizing an ample space for the social analysis of the current modes of inscription of the "I." (Fig. 1.2).

Thus, by using a methodological triad that theoretically associates a media archaeology (Parikka, 2012) with the experience (Foucault, 2011a) of a sensory ethnography[6] (Pink, 2009), formally organized through a *vampyroteuthic* game of analogies that complement one another like an inverted mirror (Bernardo, 2011), in the intersection between mediation and meditation we will seek to give voice to Flusser's call for a "phenomenology of the body" (Flusser, 1974, p. 2). A phenomenology that, in the impetus of grasping the abstraction modes of the concrete experience of what is real, executes a radical dive: a dive into both the collective unconscious exteriorized on the screen as well as into the individual unconscious interiorized in the mind. For this reason, less than confessions, using the first-person narrative aims to launch provocations. Provocations that invite the reader-accomplice to get implicated as a co-author in the current games of self-narrative mediation: as Didi-Huberman would put (2012,

[6] In this sense, the sensory ethnography proposed by Pink (2009) fits very well with the goal of a phenomenology of the body that takes the researcher-body as a research experiment since it suggests a new relationship between theory and practice. Instead of focusing only on representational or linguistic aspects, it is a question of rethinking ethnographic methods with attention to phenomenology, that is, perception, experience, and sensory categories. Thus, as in the construction of knowledge there is always the interference of meanings in the interaction process, ethnographic practice is now understood as an enterprise that "embraces our multisensory bodily involvement with others (perhaps through participation in activities, or exploring their understandings in part verbally) and with their social, material, discursive, and sensory environments (PINK, 2009, p. 4/25).

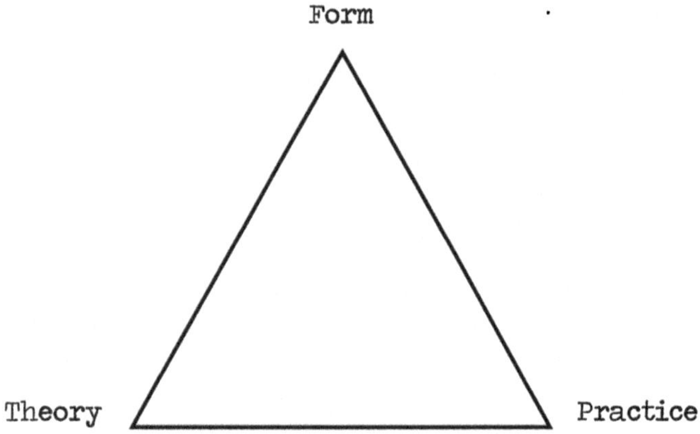

Fig. 1.2 The methodological triad, by the author

p. 15) "for us to know, it is necessary to imagine ourselves." It is in this process of self-imagining and self-implication as the opening condition for the otherness of knowledge that the "I," as an argumentative tool, aims to shed a light on the experience of any other "I" that becomes seduced by the irresistible desire to go beyond the physical presence of the body: "I" becomes "you" as much as "my" body also becomes "your" body.

An anonymous body that, influenced by the processes of "saturation of the perceptive system in the new metropolitan environment" (Felinto & Pereira, 2005, p. 86), is here presented as a first medium (Felinto & Pereira; Pereira, 2006) and as a "fundamental object of the epistemological proposition that is intended within the field of Communication Studies" (Felinto & Pereira, 2005, pp. 88–89). A body that becomes word and is materialized through the relationship between screen and mind, and its respective impacts on the body on a phenomenological, sensorial, and cognitive level: affections, desires, feelings directly linked to an elaborate system of glands, hormone secretions, neural synapses, and energy vortexes or "*cakras.*" That because, as the surface of events inscription[7], "the

[7] It is worth mentioning that Foucault (2011b) understands event as being "not a decision, a treaty, a realm, or a battle, but a relationship of forces that is inverted, a confiscated power, a vocabulary reclaimed and turned against its users, a domination that is weakened, distended, poisoned, and yet another kind that makes its entrance under a mask" (p. 28). Thus, the author presents the events as an immanent instance concerning the forces at play that create conflicts and disputes in the exercise of power.

body—and everything that touches it: diet, climate and soil—is the place of the *Herkunft*. The body manifests the stigmata of past experience and also gives rise do desires, failings and errors. (Foucault, 2011b, p. 22). Therefore, instead of working upon a prearranged concept of "affection," we will follow the so-called Affective Turn (Clough & Halley, 2007), which deals with *the matter of the matter* in order to think how the self-organization of bodies entail "bodily capacities to affect and be affected or the augmentation or decrease of a body's capacity to act, to engage, to connect" (Clough, 2008, p. 2). Capacities that, as proposed by Hardt (2015), "besides obviously calling attention to the body and its emotions" (p. 1), produce an important development: "the challenge of the affective visage primarily lies in the body/mind synthesis required by such perspective, since it would involve both reason and passion" (idem) in "a new ontology of what it means to be the human" (p. 8).

Finally, by oscillating the word in its function, at the same time, analytical and performatic, representative, and fictional, the reader—whether it being academic or non-academic—is incited to occupy both the spectator and participant's position in this diegesis open to the other's emptiness who is always affected in a singular way. After saving, collecting, and archiving, for over four years, all kinds of posts published on the Facebook news feed. After even when seeing myself being captured by the mediatic bedazzlement that became an addiction, we will evoke the masters of the human mind—the tantric yogis—to, in the obsession to be enlightened, create sensorial and cognitive bridges connecting mind and screen, screen and mind to the desire for omnipresence that today encourages the expansion of that which is experienced as the body. We hope that in sharing this limit-experience (O'leary, 2008) we can find, in the reader, a partner who is also willing to become "different."

References

Ánandamúrti, S. (1993). *Discourses on Tantra—volume 1 and 2*. Kolkata: Ananda Marga Publications.

Ariely, D. (2010). *Predictably Irrational*. New York: Harper Perennial.

Bernardo, G. (2011). Um espelho retorcido. In V. Flusser & L. Bec (Eds.), *Vampyroteuthis infernalis*. São Paulo: Annablume.

Clough, P. (2008). The Affective Turn: Political Economy, Biomedia and Bodies. *Theory, Culture & Society, 25*(1), 1–22.

Clough, Patricia, & Halley, Jean. (Org.) (2007). The Affective Turn: Theorizing the Social. Durham: Duke University Press.

Dagognet, François (2012). *O corpo*. Michel Jean Maurice Vincent (trad.). Rio de Janeiro.

Deleuze, Gilles (1978). *Cours Vincennes 24/01/1978*. In: Tradução: Francisco Traverso Fuchs. Disponível em: https://www.webdeleuze.com/textes/194, acesso em 31/03/2018.

Didi-Huberman, G. (2012). *Imagens apesar de tudo*. Lisboa: KKYM.

Felinto, E., & Santaella, L. (2012). *O explorador de abismos: Vilém Flusser e o pós-humanismo*. São Paulo: Paulus.

Felinto, E., & Pereira, V. A. (2005). A Vida dos Objetos: um Diálogo com o Pensamento da Materialidade da Comunicação. *Contemporânea: Revista de Comunicação e Cultura, Salvador, 3*(1), 75–94.

Fischer, Rosa Maria Bueno. (2001). Foucault e a análise do discurso em educação. In: *Caderno de Pesquisa*, São Paulo, n. 114.

Flusser, Vilém. (28/05/1974). unpublished correspondence, Vilém Flusser Archive, Berlin (reference number: CORRESP. 9, 6-DORA (3126).

Flusser, V. (2008). *O Universo das Imagens Técnicas—Elogio da Superficialidade*. São Paulo: Annablume.

Flusser, V. (2010). *A história do Diabo*. São Paulo: Annablume.

Flusser, V. (2011). *A dúvida*. São Paulo: Annablume.

Flusser, V., & Bec, L. (2011). *Vampyroteuthis infernalis*. São Paulo: Annablume.

Flusser, V. (2017). *O último juízo: gerações*. São Paulo: É Realizações.

Foucault, M. (1994). *Dits et Écrits—Vol I, II, III e IV*. Paris: Gallimard.

Foucault, Michel (2010a). *História da Sexualidade I:* A vontade de saber. Tradução Maria Thereza da Costa Albuquerque e J. A. Guilhon Albuquerque. São Paulo: Graal.

Foucault, M. (2010b). O sujeito e o poder. Tradução de Vera Portocarrero e Gilda Gomes Carneiro. 2 ed. In H. Dreyfus & P. Rabinow (Eds.), *Michel Foucault, uma trajetória filosófica—para além do estruturalismo e da hermenêutica* (pp. 273–295). Rio de Janeiro: Forense Universitária.

Foucault, M. (2010c). *A arqueologia do Saber. Tradução Luiz Felipe Baeta Neves* (7th ed.). Rio de Janeiro: Forense Universitária.

Foucault, M. (2011a). *O governo de si e dos outros. Tradução de Eduardo Brandão.* São Paulo: Martins Fontes.

Foucault, Michel (2011b). *Microfísica do Poder*. Machado, R. (Org). Tradução Roberto Machado. São Paulo: Graal.

Grusin, R. (2010). *Premediation: Affect and Mediality After 9/11*. New York: Palgrave Macmillan.

Hardt, Michael. (2015). Para que servem os afetos? Luiz Roberto Leite Farias (Trad.). In: *Intersemiose*, Ano IV, n. 07, Jan /Jun 2015.

Hayles, K. (2006). The Traumas of Code. *Critical Inquiry, 33*, 136–157.

Ludueña Romandini, Fabián. (2012). *A comunidade dos espectros*. I. Antropotecnia. Alexandre Nodari e Leonardo D'Ávila de Oliveira (trads.). Desterro, Florianópolis: Cultura e Barbárie.
Mlodinow, L. (2008). *The Drunkard's Walk: How Randomness Rules Our Lives*. New York: Pantheon Books.
Mlodinow, L. (2012). *Subliminal: How Your Unconscious Mind Rules Your Behavior*. New York: Pantheon Books.
O'leary, T. (2008). Foucault, Experience, Literature. *Foucault Studies, 5*, 5–25.
Parikka, J. (2012). *What Is Media Archaeology?* Cambridge: Polity Press.
Pepperell, R. (2006). Where's the Screen? The Paradoxical Relationship Between Mind and World. In R. Pepperel & M. Punt (Eds.), *Screen Consciousness: Cinema, Mind and World*. Amsterdan/New York: Rodopi.
Pereira, Vinícius Andrade (2006). Reflexões sobre as materialidadesdos meios: embodiment, afetividade e sensorialidade nas dinâmicas de comunicação das novas mídias. In: *Revista Fronteiras*—estudos midiáticos (Unisinos), v. VIII (2), p. 93-101, maio/agosto.
Pink, S. (2009). Situating Sensory Ethnography: From Academia to Intervention. In *In: Doing Sensory Ethnography*. London: Sage Publications.
Shusterman, R. (1998). *Vivendo a arte: o pensamento pragmatista e a estética popular*. São Paulo: Editora 34.
Thrift, N. (2004). Remembering the Technological Unconscious by Foregrounding Knowledges of Position. *Environment and Planning D: Society and Space, 22*(1), 175–190.

CHAPTER 2

Mediation and the Process of Outer Expansion

January of 2014. For the first time, I feel a touchscreen surface. Prior to that, my cell phone had keys, plastic buttons that I would press so that it would show me on its screen the commands I had entered. It was one of those "dumb" phones: it could not download Apps, it did not access the Internet and, to send text messages, I had to press the same buttons many times in order to transform alphabet letters, which were displayed together with numbers and symbols, into words. My day was clearly divided between offline and online time. If somebody sent me an e-mail or a message on social media, I would answer the person only after I got home and turned my computer on. Not anymore, though. With the advent of a device that offered me only a screen as an interface, I had to learn how to navigate a whole new universe of buttons that would become images to be touched. Free from having to repeatedly press the same buttons over and over again, I would now smoothly glide my fingers on a crystal that obeyed me and immediately fulfilled my demands, no longer restricted to calls and text messages. Both my gestures and my communication rhythm were no longer the same: from then on I could configure my "smart" device to receive e-mails in real time and I started to access social media via "Apps" that would insistently tell me the number of status updates added to my profile at the very same moment they would occur.

Almost as if by osmosis, without a significant and conscious moment of decision, mediation came into my life. Subtle osmosis, like the advertising-related osmosis that would not exactly force me to comply with its

C. Mozzini-Alister, *Does Social Media Have Limits?*,
https://doi.org/10.1007/978-3-030-74120-4_2

imperatives of happiness by saying: "buy a phone with a photosensitive screen, access to the internet and start downloading social media Apps now!" It was only inciting me,[1] seducing me, sweetly whispering to me in every outdoor, television commercial, newspaper ad, product package, YouTube video, and piece of advertisement generated based on my digital traces[2]: "you can be happier, you can be more than your body, you can be limitlessness if you are connected to what we are offering you: but the choice is all yours." I wanted to "be connected." I wanted to be more than my body. I wanted to be unlimited. I was afraid "to be left behind" and feel like a fish out of water. And seeing that "everybody" around me already knew what was going on, I also wanted to be a "part of it." I was tired of having to respond to the ironic comments about how my "less-than-smart" phone was obsolete and, as a consequence, tired of feeling like an "outsider," "isolated," "disconnected" from something that so many people were so seemingly fascinated to share. Albeit suspicious about it at first, I unwillingly gave in. I gave in to the advertising imperatives and to the social pressure and all of a sudden there I was, using my fingers to explore a slippery screen that was opening the gates of Facebook and its App to me.

I was initiated into Facebook in 2009, during an academic exchange at the National University of Cordoba, in Argentina. While to me Facebook sounded like the "latest shiny new thing," the foreigners with whom I became friends in this period had already been surfing that "new wave" for

[1] It is important to stress that such presence, even though continuous, does not entail, in any way, a kind of passive obedience to the incitements of advertisement since "the effects of advertisement upon the modes of being are not produced because it forces people to do something that, deep down, they do not desire. Such effects belong to the order of incitement, since the discourse related to advertisement presents ways of thinking, feeling, having fun, taking care of oneself, consuming, and so on, evidently linking them to products and services" (Hennigen, 2011, p. 5).

[2] Such strategies of utilizing web protocols in order to create personalized ads can be seen not only on social networks but also on research websites and on e-mail services such as Google and Gmail since "the monitoring of information and actions belonging to individuals on the cyberspace is intrinsic to any search engine, and it is inherent to its functioning and efficiency" (Bruno, 2010, p. 158). In this sense, a digital trace can not only be understood as a "vestige of an action performed by any given individual in the cyberspace" (Bruno, 2012, p. 5), but also by automatized processes, performed by robots and meta-robots through the sophisticated programming of algorithms. Therefore, there is no way of being online without leaving traces—unless one takes certain measures to avoid that—because they follow any performed action due to the fact that "the more one desires to inscribe oneself into the network, the more one will leave involuntary traces behind" (p. 7).

a while. I was an "Orkut" user, a social network created by Google that became a fever in Brazil. Due to my naivety, I thought Orkut was "the" social media bringing people from all around the world together. But again, no: my trendy foreign friends were already using the website created in 2004 by Mark Zuckerberg. And since I wanted to keep in touch with those friends when they returned home, I created another profile on a new social network. At first, I did not quite understand how it worked or what were the so-called timeline and newsfeed. But it did not take long for that new architecture to become familiar to me and, besides contacting the friends I had made in Argentina, I was also contacting my high school and college friends who slowly were migrating from Orkut to Facebook. Such stampede was so fierce that Orkut was no longer a fever: lukewarm and poorly accessed, it was extinguished in 2014 in tandem with my first finger-related forays onto screens tuned to the Facebook wave of connection.

Since then, the action of mediating took me to another level of connectivity: I could do more than just keeping in touch with those who were no longer close to me, reconvening with childhood friends and adding people whom I would meet during my daily life. I was doing all that by simply gliding my fingertips on a smooth and slick surface, that seemed to caress me. More than just pressing buttons on a mouse or keys on a keyboard when I got home, by remotely logging into my profile through the Facebook "App" on a mobile device I was also giving my fingertips a new power: with such a small and quick gesture, the magical possibility of being close to everyone who was far away from me became real. By transforming the world into an accessible miniature (Benjamin, 1985) to the gliding of my fingers, that device and its crystal interface full of images started to exert over me a tremendous attraction. I wanted to merge with the screen, unite myself to the network, be one with everything. It was hard to control the urge to be there, always accessible and with free access to everyone with whom I have crossed paths in life: the distinct feeling that a plethora of things were happening while I was disconnected to eat, sleep, read, write, go out, and meditate, made me do all of those things in parallel, with my phone lying right next to me.

Little by little, I started desiring more and more to be present in that space that would take me beyond the space occupied by my body—and it seems I was not the only one: according to the Kantar Institute, world leader in statistical measurements, the number of times users between 17 and 25 years of age in the United States unlock their iPhone screens to check on their devices is, on average, 123 per day. In other words, around eight times for every waking hour, taking into account sleeping hours and without considering the amount of time people would use their phones after unlocking it. "If you are a young adult and have an iPhone, the only physical activities that you can perform more times than checking your phone are breathing, swallowing saliva, and blinking," says a news article on *Portal UOL Tecnologia*.[3] In Brazil, for instance, another study conducted by the Global Mobile Consumer Survey pointed out that the average number of times a phone is unlocked per day is 78, and that "in the first five minutes after waking up, 57% of Brazilians already have a phone in their hands",[4] ready to perform three main activities: taking pictures (67%), accessing social networks (55%), and watching short videos (40%). In a nutshell: touching images. What is it that we are looking for on the screen? How does this uncontrollable desire for mediation provide clues on the modes of expansion of the biological substratum? And in what way does the readily instantaneous process of mediation manifest the desire for omnipresence?

By facing these questions, we will start our journey toward the first of the anthropotechnics to be studied: mediation. Inspired by the phenomenological philosophy of the image proposed by Vilém Flusser, we will investigate the desire for limitlessness through smartphone screens tuned into social media Apps, more specifically, to Facebook. The reason for this choice is related to the fact that today Facebook is the oldest living platform with the largest number of users whose openly declared mission is to encompass every single person in the world. Even though nowadays there are so many different social networks such as LinkedIn, Reddit, Instagram, YouTube, TikTok, and Reels—and the incredible incorporation of Snapchat's logic of "filters" and "stories" into all these platforms—we will now focus on Facebook as a primary model, a matrix to unfold the subtle relationships between screen and mind, outer and inner expansion beyond the physical presence of the body.

Therefore, while mapping the desire for connection as a means of connection to the divine touch, we will conduct a material archaeology of the

[3] Complete article available at: https://tecnologia.uol.com.br/noticias/redacao/2015/10/04/jovens-adultos-checam-iphone-em-media-123-vezes-por-dia-diz--pesquisa.htm, accessed on 07/27/2020.

[4] Complete article available at: https://odia.ig.com.br/_conteudo/noticia/brasil/2015-10-28/mais-da-metade-dos-brasileiros-usam-celular-cinco-minutos-depois-de--acordar.html, accessed on 07/27/2020.

screen across three stages: in the first, we will study the process of energetic symbiosis of the shock through the touch between fingers and planes that produces the device as a body; after that, we will map the activation path of the device's body via the duplication of oneself as a profile dot; finally, we will make visible the processes of irradiation between person and profile, body and device's body based on social interaction methods that aim to merge the individual's mind with the Great Mind of Facebook. We hope that, with such endeavor, we can shine a light on the modes of convergence between body and light that arise from the study of the current mechanisms of the body's electrocution.

2.1 Screens and Shocks

The decisive factor in decipheringimages has to do with planes. (Vilém Flusser)

March of 2013. I was finally finishing my Research Master's Degree. Those were two intense years of hard work, blindly fumbling with the study of how advertising was establishing the digital connection as a new universal human right: I had become a fierce critic of the happiness imperatives diluted in ads promising a better life—as long as such a life was a connected one (Mozzini-Alister, 2019). A "connected life," that is, a life that feasts on the thousand possibilities offered by the digital connection through a smartphone with touchscreen technology. I still did not have that magical portable device capable of accessing wireless frequencies that promised to "connect me to the world" at the speed of "the blink of an eye." I was part of the 64% who, in Brazil, were still not properly incited to the digital connection and would not only use "basic phones"[5] but would also access social media only from their homes. Perhaps that is the reason why it was easier to adopt a critical stance and point out the regimes of truth involved in the internalization of the advertisement's discourse: I was at a safe and shielded distance from both smartphones and Facebook. After having turned the policy of use of such network from the inside out in order to write my dissertation, I considered absolutely pathetic the way in which we would gratuitously work for said network by posting smiling

[5] Data extracted from the article found at: http://tecnologia.ig.com.br/2013-01-18/entre-os-celulares-usados-no-brasil-36-sao-smartphones-diz-nielsen.html. Accessed on 07/27/2020.

pictures followed by beautiful phrases just to earn a few likes and a certain social status.

However, as the wisdom of the saying goes, "what goes around, comes around" and, bitterly, I had to swallow that feeling of superiority disguised as a critical stance. Tired of having to thoroughly inspect Facebook in order to find out what advertisements would produce as empirical materials, I decisively deactivated my profile for an undetermined amount of time. However, an unexpected reaction once again brought forth that network to the center of my affections. Baffled by the disappearance of my profile, my partner at the time uttered a blunt statement: "It felt like death." For two years and a half we were having a long-distance relationship and now that we were finally together, he says this piece of "nonsense": why would he long for something that is right next to him? Not even the erudite doctor in Philosophy, who got his degree in the United States, could avoid feeling bewildered by a strange feeling of mourning. A few months after this episode, already back to said social media, our relationship of seven years came to an end. To see all the pictures we had together being drastically deleted caused me an intense bout of crying. I was in shock: with the unravelling of my love plans and the accumulated tiredness due to the stress of my Master's studies, came the news that I had not been approved for the PhD program in Communication at UERJ, that my work as an art educator at the Mercosul Biennial had ended, that I had to say goodbye to the eight years of living in Porto Alegre, and that I had to go back to João Pessoa to live with my parents after struggling so hard to earn my independence. Everything that felt stable, all of a sudden came crashing down. I was not hungry or thirsty, I did not have a terminal disease, and I was not homeless, but there was a hole inside my chest. A hole deep enough to turn my life into an uncontrollable sea of tears: overwhelmed by the despair of not feeling myself in my own body, I became another number in the statistics of depression in the world.

It was at that moment that I put my hands on my first smartphone. The feeling of not being adequate as a person allied with the feeling of technological insufficiency started to bother me: I was still accessing all those smiling profiles on Facebook by pressing the buttons of a mouse connected to a computer screen. Ironically, I also wanted to experience what advertisement was marketing as "our form of infinity" (Vaz, 2010, p. 201), which artists such as Roy Ascott (2010) would consider to be the space where "we can truly fly" (p. 245) and that everyone seemed to be enjoying in selfies shared in real time. No longer at rock bottom, "reconnected" to myself after spending a week immersed in a miraculous meditation retreat, I also wanted to digitally "connect myself" to the world and leave behind once and for all that "backwards" Camila who was drowning in tears. And after the retreat, life was only getting better: the doctorate program that had refused me, invited me at the last minute. Somebody had given up, opening the way for me to be accepted. As a Ph.D. student, living in the coveted "Marvelous City,"[6] free of any existential crisis and possessing a brand new smartphone, now I was longing to share on the network I so fiercely condemned the fortuitous lucky strike that would lead me back to happiness. However, by losing myself in this impetus for extroversion, unbeknownst to me, I embraced the first vector of expansion and deconstruction of the physical form of the biological substratum: the outer expansion.

What are the technical proceedings of this mode of expansion? To face such question, it is necessary to keep working on what Flusser pointed out as the deciphering of planes. Planes, that is, screens, mnemonic supports, surfaces for the image's material inscription. Depthless structures, bidimensional, possessing only height and width. Planes that are manifested today in the form of smartphone screens capable of "connecting us to the world" through the flickering of images that demand only one thing: that we also connect the energy of our fingertips to the energy of the device's electronic circuits through a shock. A subtle, instantaneous, automatic shock.[7] But necessary if we want to penetrate the magical kingdom reserved to the social sharing of happiness called "Facebook." It might be hard to perceive any dimension of the process in gliding fingers on screens since it is such a simple, quite fleeting, and immediate gesture. Without much effort, by the minimal moving of fingers, today we make Narcissus's repressed dream come true: to see the sharpness of our own image without the risk of drowning as we are trying to grasp it. But we also added a new feature of finesse to it: the desire to grasp our own image combines itself with the desire to remotely connect ourselves to the images of "others" scattered across houses, streets, cities, states, countries, continents,

[6] Marvelous City is how Brazilians call the city of Rio de Janeiro.

[7] The concept of shock, in the nineteenth century, was related to the training of bodies living in the city by means of the speed of images seeing in movies and the impetuous fun of amusement parks (Crary, 2010). However, today this shock is no longer psychological. It became literal: touch and shock are inescapably intertwined in the practices of mediation due to the ascension of the touchscreen technology as a trivial, banal, and recurrent gesture.

and all over the globe through "social networks."[8] What would be the techniques that allow us to put an end to the distances that separate the connection of my individual "Self" from all the other "Knots/Us[9]" of the network[10]?

Flusser (2008) would say that, by facing the urge of becoming conscious of the world's concreteness and moving away from it in order to grasp it better, we became so distant that we no longer grasp things, but intermediates of things. In other words, instead of creating a circle by

[8] It is important to stress that the notion of social network predates the existence of any website like Orkut, Facebook, Instagram, Reddit, or Twitter because what is understood today as social networks was already existing even before the emergence of the Internet. Based on the anthropological and sociological studies from the 1950s, scholars like J. A. Barnes, J. C. Mitchell, P. Bourdieu, and J. S. Coleman exhaustively worked on the most diverse approaches related to such concept. In general, a common core among several theories considers social networks as being a study about the processes of belonging to certain groups or categories. Social networks would be, then, directly associated to the accumulation of social capital, that is, to the individual advantages and benefits that are obtained via one's insertion into these communities of living and participation. Nevertheless, since the ascension of the web and the explosion, during the 2000s, of a series of sites also called social networks, the creation of social capital assumed new outlines because "the internet, many times, represents an alternative way to the participation in social groups. Mediation through the computer, thus, would be a way of building social capital, allowing individuals to access other networks and groups" (Recuero, 2009, p. 52).

[9] In Portuguese, "Nós" means both "Us" (we) and "Knots". There is a word play here between Nós (Us) and Nós (Knots) on the web.

[10] There are many scholars who identify the notion of network with "an epistemological receiver or crystallizer" (Musso, 2010, p. 17), an "empirical character on the ontology of the present" (Kastrup, 2010, p. 88), a kind of "paradigm and main character pertaining to current changes" (Parente, 2010, p. 92), a "network-space, dynamic, infinite, configured in the dynamics of associations" (Lemos, 2012, p. 2). Historically, the notion of network is quite old: in the East, there is the ancient legacy of a holistic way of thinking that conceives the interconnection between life and cosmos as an inseparable unit. In the Western context, Musso (2010) points out three main movements: the first would be the idea of network as an object external to the body, exemplified by the French concept of ***réseau*** such as fishing and hunting nets and fabrics that enwrap the body. The moment these networks enter the structures and fabrics of the body and appear in nature as crystals and minerals ends up creating a second movement. The third one would be the rupture that engendered the notion of network in the passage between the eighteenth and the nineteenth centuries, when the network is no longer related to the body and to nature, and presents itself as something that can be artificially construed in relation to the space. The exteriorization process of such networks as technical artifacts is even more intensified after the emergence of the Internet and its expansion beyond its restricted use by the military, and becomes articulated as an organizational platform that traverses the capitalistic model.

uniting the extremities of a thread to get closer to it, we write the word "circle," designate such gathering of letters as a phoneme to be uttered and, ultimately, we simply draw said shape on a plane. *Voilà*: we brought a circle closer to us without having to have it in our hands. People-wise, it works in the same way. When we want to touch someone without being close to them, we would single them out by writing a "proper name," assigning to the name a particular sequence of sounds that can be uttered and, finally, translating the physical outlines of that being endowed with a singular name into bidimensional images. It is no coincidence that Emanuele Coccia proposes that for an image to be born we only need to have "a separation of the form of a thing from its relationship to the place of its existence: *where the form is out of place, an image happens*" (Coccia, 2010, p. 23). And it seems that today more and more people are exercising this dissociation between form and place, body and space occupied by the body, in order to connect themselves to images of social media profiles every day: "We are making progress in connecting the world, and now we are going to bring the world even closer together,"[11] celebrated Mark Zuckerberg in 2017, when Facebook reached two billion users.

It is not by chance that this number grows in parallel to the access expansion to said social media through smartphone Apps. In 2013, for the first time smartphones were sold more than dumbphones or computers.[12] Today the number of "*smartphones*" since they are "digitally connected" equaled the number of humans on the planet. And even though such a high number exists due to the fact that some people have more than just one device, in the same way as the five billion inscriptions in mobile phone services[13] reflect the existence of more than one SIM card or chip per user, these results are still astounding. With China, India, and the United States on the top three positions concerning the number of phones being used, Brazil comes right after in fourth place. With around 234.6 million of wireless connections, the country has equaled its number of inhabitants to the number of smartphones since 2017. Therefore, to increase the penetration index in rural places, small towns, peripheral, poor, and socially vulnerable areas is a crucial goal to make possible for each individual to

[11] Complete article at: https://g1.globo.com/tecnologia/noticia/facebook-atinge-os-2-bilhoes-de-usuarios.ghtml. Accessed on 07/27/2020.

[12] Data available at: https://www.businessinsider.com.au/smartphones-and-tablets-penetrate-globe-2013-10. Accessed on 07/27/2020.

[13] Full report at: https://www.gsma.com/newsroom/press-release/number-mobile-subscribers-worldwide-hits-5-billion/. Accessed on 07/27/2020.

access this virtual agora of sociability that promises to make the world happier because less distant. Due to this fact, the African continent is the one with the fastest expansion in the number of these devices: even though many times there is a lack of basic sanitation, the possibility of going beyond the "basic" and becoming connected to what takes us "beyond" the reach of the corporeal substratum must become the norm, regardless of anything else.

"Beyond," that is, no longer conditioned by the limited reach of arms, legs, vision, voice, ears, sense of smell and touch. How we wish we could make that happen by stretching the circle's thread and adding two open and lidless cans to its extremities. When I was a kid, we would play a game called "wireless phone" and it worked just fine. Today, however, the desire of wanting to be more than our body and to connect ourselves on a global scale is modulated by the massive use of screens on smart devices photosensitive to touch. Thus, the first and crude reality that one has to face before entering the gates that lead to Facebook is to amass money to acquire one of these devices. One has also to link it to a mobile phone carrier capable of endowing said device with an unrepeatable and unique number, one that identifies and enables it to communicate with all the remaining phones, whether they are a land line, mobile, smart, or dumb. And it is almost illogical to have a smartphone that cannot digitally connect. Hence, we turn once again to phone carriers in order to acquire a bundle of data services known as "4G," "5G," or "6G." By adding this very profitable service to the functionality of a device, the first stages of outer expansion are fulfilled—the financial and the bureaucratic ones. Tied to a variable monthly fee, finally the "self" acquires the object of consumption that today materializes the desire for connection.

The possession of the new device causes a certain euphoria: what novelty can I perform with it? How is my life going to change now that I have it and can simultaneously share my affections on a platform interconnected on a global scale? For somebody who, like me, was acquiring their first smartphone in the beginning of 2014, such feat implied relearning an entire code, a series of etiquette rules so that an actual relationship between the device and myself could take place. And the very first rule that we have to come to terms with is the absence of a keyboard: how can we make this thing work? In this moment, it is necessary to employ the faculty of intuition: under the name "ubiquitous computing," since 1988, scholars like Mark Weiser have been working to develop intuitive interfaces capable of being connected to a network. In order to provide experiences that are no longer limited by the static structure of desktop computers and that can be accessed at anytime and anywhere, Mark came up with four premises: (a) the purpose of a computer is to help you do something else; (b) the best computer is quiet, invisible, and servant; (c) technology must be a continuation of the human unconscious, that is, the more the users can perform any task simply by using their intuition, the more intelligent is the user; and, finally, (d) technology should create calm, it has to inform without demanding our focus or attention. With one detail, though: what he would consider a "computer," today is a "telephone" that became a "smartphone": mobile, portable, and connected.

Thus, without much effort or reflection, only by intuition, the act of pressing the residual "power" button on the device's side wakes it up and offers to the eyes images to be pleasantly touched. One only needs to touch such images and, submissively, the smooth plane obeys without hesitation: finger and screen are energetically connected. How does this command, so immediate that it seems to not have a process, happened? Would this be one of the reasons for the *frisson* surrounding the current practices of mediation? By satisfying the desire for connection, a subtle desire is aroused in those who possess a smartphone: the desire to technically penetrate inside the device's circuitry through the shock of the touch on the screen. Would it be possible to become one with everything without first becoming one with the device? For that to happen, it is necessary to let ourselves go and be affected by this new kind of skin made of three thin layers: (a) one made of a ultra-resistant glass; (b) one made of a transparent conducting film; and finally, (c) one made of symbiotic capacitors. Ingenious skin. Its creation and combination are the result of a true *eureka* of engineering, starting with the crystal: on the outermost layer, a new kind of glass quite distinct from a window glass. A transgenic glass.[14] "Transgenic" because its molecular structure is genetically altered with the goal of producing an ultra-resistant version: vigorous enough to withstand falls, scratches, and impacts but, at the same time, narrow enough to allow the screen to be sensitive to the touch. And why is it so indispensable to turn the glass into a super-glass?

[14] Probably developed by a company called Corning, world leader and holder of the monopoly on glass, ceramics, and optics research, which on its website pompously announces the advent of the "Glass Age": "throughout history, materials transformed society and culture. There were the Age of Stone, the Age of Bronze, and the Age of Steel. This is the Age of Glass." Company website available at: http://www.corning.com/worldwide/en.html. Accessed on 07/27/2020.

More than just a matter of the product's durability and smoothness, a resilient glass is crucial to protect and create conditions for the activation of what is immediately below: the "transparent conducting film" (TCF). TCFs became mandatory items on devices that possess technologies such as liquid crystal, LED, photovoltaic lighting, and touchscreen features. Today, the most common composite made for such use is called "indium tin oxide" or ITO.[15] Its existence is the result of the combination between indium and titanium ores extracted under predatory and exploratory mining conditions and industrially arranged in the form of a solid that behaves as a liquid: at the same time translucent and capable of conducting electricity. If for a long time, conducting and non-conducting materials would be related to either non-transparent metals capable of conducting current or to translucent materials unable to conduct electricity, it is in the passage from the twentieth to the twenty-first centuries and the emergence of TCFs that this situation assumes a new dimension. And such change is in tune with a new discovery by the engineering sciences: the creation of symbiotic capacitors.

Right below the TFC, in the depths of the surface, it is where the capacitors can be found: electronic terminals responsible for temporarily storing energy. Capable of creating an elaborated circuit that emits and circulates photons and electrons, their mechanism is similar to a battery: one has to place two conducting materials separated by another non-conducting material, which will in turn produce a potential difference (PD) or tension. With two small details, though. First: whereas the battery feeds the device through the slow liberation of its energy, the capacitor instantly releases all of its stored energy like the flash of a camera. And the second one: on surfaces sensitive to touch, instead of two, there is only one slab of conducting material. The other slab is the body itself, that is, the fingertips that act as proficient electricity conductors. Hence, we have "symbiotic capacitors" that gave the name to the current "capacitive screens." Screens that during the timid interaction between the electromagnetic fields of fingers and capacitors, extract a minimal fraction of our

[15] Pertaining to ITO, tin is an abundant and accessible metal, with a silvery color, malleable and solid in ambient conditions, easily oxidized by air and resistant to corrosion. Indium is malleable, smooth, glossy, highly toxic, and today emerges as the centerpiece in the current "gold rush" due to its scarcity and, therefore, its economic relevance. As a result, other, and less expensive, mixtures of metals combined into oxide forms are also created as an alternative to ITOs, such as the aluminum-doped zinc-oxide (AZO) and the indium-doped cadmium-oxide.

own energy in order to activate the device's electronic circuits. Instead of the inert material of a pencil, pen, or brush, it is the soft and electrified tissue of our skin that ignites the commands programmed to govern electrons, photons, and bits in the production of images. In other words, only when we literally donate our own energy to the device and connect ourselves to it, its capacitors charge, release, and automatically give back to us images to be grasped by a shock. What kind of skin are we touching?

Skin without pores, hair, or sweat. However, still a sensitive skin. Smooth: absolutely smooth, the skin that wraps the current smartphones' exoskeleton exhibits images and caresses us by gently electrocuting those who touch it. If we had eyes on the tip of our fingers, and infinitely stretching arms, we would not need to expand our sensory-motor apparatus through sensors connected to smart devices: at last, we would fulfil the dream of wrapping our arms around the universe and hold it. But no. The eyes are located on the head. The fingers, on the hands. And the arms cannot stretch at all. What is left for us is the act of submitting to the device's programmed physiology, electromagnetically entering its circuits through a shock. An almost imperceptible shock. However, it is powerful enough since it is laden and extensive to the billions of individuals scattered across the four corners of the planet, with their phones ready to connect them to any other corner. What happens when the smoothness of the human skin meets the smoothness of the device's skin? Where does this nanoshock that the device and I intimately exchange go to? Where in the body does it enters, and where in the device does it penetrates? The exact path is unknown, but we can feel its effects: it directly affects the nervous system and captures the touch by the shock as a condition for the very act of seeing[16] (Deleuze, 2007). In this superficial and invisible exchange of electric fluids, a game of mutual excitement is inaugurated: we gave in to the shock, giving to it a small amount of our energy through the touch of fingertips, and the screen immediately returns such surrendering gesture

[16]The current fusion between the senses of vision and touch in order to produce tactile images and imagetic touches can be thought via Deleuze (2007) based on the concept of haptic. Differently from something optic, that is, of something strictly related to the vision, something haptic is that which promotes a rupture from the logical, representative, and perspectival regime of a world organized by the vision while engendering a presence. A presence capable of creation chunks of feelings, of touching the nervous system and generating a space that loses its extensive and measurable depth. In the haptic regime, the eye assumes a tactile function: from a spatial area of proximities, the vision acts like a tactile proof of what is constituted as presence.

with the flickering of images to be touched yet again. I touch it. I handle it. I glide it. I caress it. I feel pleasure. With every new touch, a desire is quenched whereas another one is incited.

We only have to rest our fingers and stop touching it for the screen's skin to literally turn off. The device that looked so smart becomes a mere inanimate object, a black and thin imageless rectangle. The screen is dead. Only through the touch of my fingers it can be resurrected. Such control is seducing. And incites us to always want more: more and more images to be caressed by the epidermal tissue. Then, I smear my fingers going up and down and sideways, from left to right, in the corners and diagonals on this crystal skin that passively feeds off my energy in order to respond to my most wild movements. I acquire the power to, simply by the touch of my fingers, give life to the device. Without realizing, I feast on the pleasure arising from the obedience of a tissue that satisfies, in a personalized way, all of my desires. The screen and I, carbon skin and crystal skin, establish a relationship not only intimate but also erotic. We oscillate between close and distant, "between control and renunciation, between being the giver and being the receiver" (Marks, 2002, p. 16). And in this ecstasy of experiencing the sense of self and of self-control being at the same time removed and restored (Marks, 2002), we create a strong energetic bond. The smartphone become a part and parcel of living, of our daily habits, of sleeping and waking up, of moving and stopping, of the most private and public of affections: they not only rule us but they also feed us. Would that be the reason why so many bars and restaurants also serve "free Wi-Fi" as part of their menu? Thus, by exposing its circuits, and leaving them open to a fullness that arises only through the activating powers of my touch, the device stops being just a device and becomes a body. A body that, with a few touches on the "Play/Apple/Whatever Store" installs on the screen a new icon: the "Facebook" App ready to provide access to new layers of that desire for the limitlessness.

2.2 Screens and Points

We are beginning to weave images using the points of the nothing outside, and to weave information networks using the data bits of the nothing inside. (Vilém Flusser)

Visualize a circle that spins around itself (Fig. 2.1). Also visualize that outside the circle there is a point. Seduced by advertisement's fruitfulness

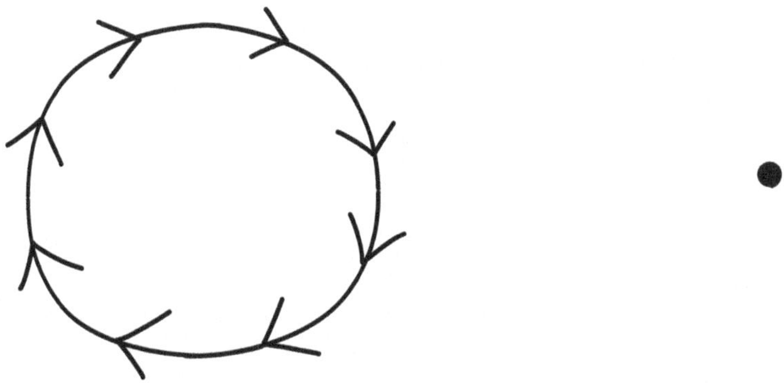

Fig. 2.1 The circle of the individual "I" and point of the profile "Self", by the author

and with the addition of social pressure, the circle wants to be a dot on the network: an individual "I" is not satisfied by living a life disconnected from the possibility of inhabiting a profile "Self" that connects it to two billion other dots on the Great Us/Knot of Facebook. How to connect that limitless point of "Self" to the circle's rotation contained in its volumetric limits? What can we do to unite the unitary "I" to the multipliable "Self"? Before the agonizing distance that separates circle and point, there is no other choice but to extrovert the tribulations of life and turn to the outside or, in other words: "to duplicate oneself by the tangent." Etymologically, the word "tangent," from the Latin "*tangens*", means "something that touches or acts along." This, in geometric terms, is represented by a straight line that lightly touches the circle on a single spot of the curve without invading the interiorities of what is inside. Avoiding the circumference's nucleus, in the mediation practices such straight line implies moving away from the center where the individual soul inhabits in order to indirectly touch it through the duplication of oneself as a profile. Indirectly, that is, by creating an external path that allows for the connection of the "I" to the totality of the network's "Us/Knots". This is the case because, properly excited by the subtle shock of the device, our fingertips now inaugurate a new phase of symbiosis: to give life to the device through the process of linking "person" and "profile" using the method of "mediation" (Fig. 2.2).

I could barely have imagined that the actions performed to get around the unpleasantness of not having a smartphone, such as amassing money

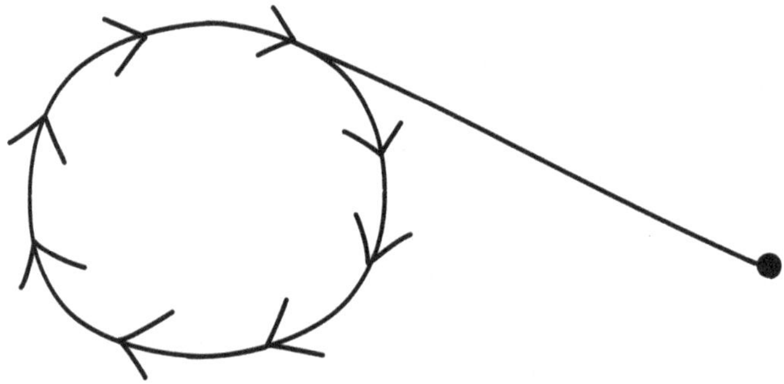

Fig. 2.2 "To duplicate oneself by the tangent", by the author

to buy it, acquiring a mobile Internet service, and learning how to navigate the keyboard through the electrocution of fingers on a screen, would be enough to complete that tangential line: with them, it was already possible to download the Facebook App. Finally, the "I" person and the "Self" profile are creating a feedback channel. But adding to the body of the device a new function without actually using it is similar to dying of thirst when there is available water: it simply doesn't make sense. Thus, now it is up to the "I" to be quickly initiated into the process of applying oneself in the App's use. "App": word possessing a double meaning that, from the Latin "*applicativus*," relates to the act of approximating, interlacing, uniting and, from the Latin "*applicare*," links the prefix "*ad*" (toward) to the root word "*plicare*" (to double), in the sense of devoting oneself to the act of putting things into contact with one another. In other words, to link the rotational movement of the circle of the "I" to the point of the profile "Self" that is right there, outside the body and within the reach of our hands, waiting to be rewarded with data, images, and personal information. This connection requires devotion. Devotion not in the sense of being a difficult or arduous job, but because it is necessary to transform the tangential line into more than a mere straight line: it must become a line of flux, recursiveness, and continuous feedback so that it can create an umbilical cord that animates, that is, that grants an "*anima*," from the Latin "soul," to the profile shown on the device's crystal skin (Fig. 2.3).

However, to establish a wide and lively two-way avenue between person and profile, first we have to open a one-way road, a step-by-step process through which the profile may be created. In this first stage, instead

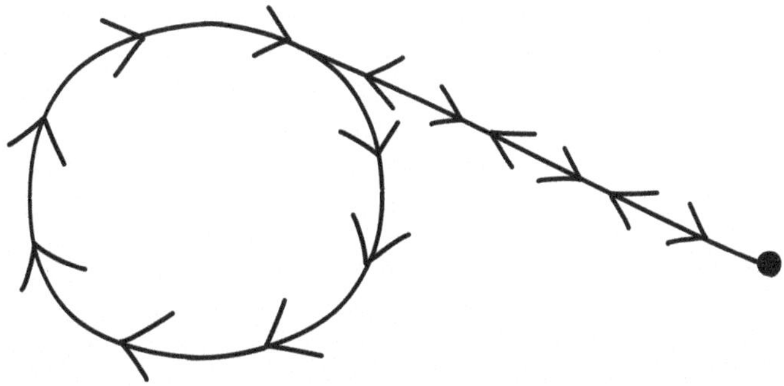

Fig. 2.3 The feedback route between circle and dot, by the author

of torturing myself with existential questions such as "where did I come from?", "where am I going?", or "who am I?", and contorting the body to meditate and reconnect ourselves to the soul's interiority, I only have to follow the commands programmed within the easy-going and friendly "App." It is the App that conducts this necessary stage of animating the smart body throughout the externalizing process of desires in order to expand the ecstasy derived from the electrical brush with the screen's skin. With the eyes open and the fingers ready, and by clicking on the blue and square icon with the "f" of Facebook, I open the first and foremost interlocution window between person and profile filled with images of hearts, laughter, and surprise emojis, photos of friends hugging, landscapes and, of course, a big thumb pointing up in the middle of everything carrying the phrase "Sign Up." To that end, the App is solicitous: "We will help you create a new account in a few simple steps." It is up to the fingertips to simply press the blue button "Next": without roundabouts, the "I" made of flesh, bones, organs, and senses advances in the creation of an "account" that authorizes the welcoming of the existence of its double, the profile's "Self".

For a long time, the term "account" was related to meanings such as: a small adorn used in the manufacturing of necklaces, any arithmetical operation, the payment for acquired goods, the registering of bank transactions and clients of an advertisement company. Today, in order to become a profile it is also necessary to create an account in which a pool of data is electronically linked to a user name. It is no accident that "account" comes from the same Latin root word of "computer": the joining of the prefix "*con*" (with, together) with the root word "*putare*" (to estimate, to think, to order), which composes the word "*computare.*" And by ordering a whole structure with the goal of organizing it methodically, the new stage to be completed is precisely to designate a name for the profile's account. In it, the question "What is your name?" offers to the fingers the act of typing letters distributed into two lines, one for the first name, another for the last name. If by any chance you try to write an unorthodox name or one that is not similar to your real name, a message in red saying "One of the requirements is that everybody who is on Facebook uses their day to day name, the name by which your friends call you" will promptly appear on the screen. Being called by the name people call you at a bar, in your family, and in daily circles of interaction by more than two billion other names is something that not only attracts but also authenticates the profile: what is the point of duplicating oneself to be just a solitary dot on the network? Thus, only after colluding with the pact to expand the socialization scope as a condition and making your real name your profile name, it is possible to, once again, hit "Next."

Automatically, a new page asks the user: "What is your birth date?" As long as one types any number above the thirteen years of age, it does not matter if the day, the month, and the year are the same as the birth date written on one's birth certificate. However, there is a caveat: if you want to have your birthday celebrated not only by your friends and close relatives, but also by your friends scattered across the network, it is better to avoid creating a random date. After all, why lie? Hence, I introduce the real birth date in order to again hit "Next." Then, the App starts to get into more intimate subjects and it wants you to be specific about them. It imposes the question "What is your gender?" with three possible answers: "female," "female," or "custom," recently inserted as a possibility. Until then, there was no place to insert something like "other" or "undisclosed," such as the Gmail registering already offered. Before this change, people with a fluid gender identity such as drag queens, transgender, genderqueer, or even asexual people were forced to avoid conflicts and choose only between "female" and "male." Thankfully, not anymore. Now, the "custom" button can be selected if you "chose another gender, or if you'd rather not say", says the subtitle bellow the option. In a historical moment where the LGBTQIA+ questions are more and more on the spotlight, only by answering this question it is possible to hit "Next" and keep going through the process of folding ourselves into a profile dot. After this page, another one asks the user for information about one's country and phone number to be utilized when accessing the App. This time it appears to be more flexible and offers to the user the option to register via one's e-mail address as well. But what if, just like the 3.23 billion people without access to the Internet in the world,[17] it is not possible to register by using any of them? Well, in this case the game is over right here.

Let us not be fatalists, though. There are still 4.57 billion people who both can and want to keep going through the process of inserting themselves via the App's use and are in good stand to advance to the next stage: to assign a password to the phone's number or e-mail registered in the profile's account. It could be any password: as long as it is secret and has at least six characters. This is enough to make the access to the profile something exclusive to the person who registered it. There is not even the need to memorize it. If the person willingly chooses so, the App can be responsible for saving the key that opens the gates of Facebook with a single touch on the screen. However, such amenity is only conceded after one has concluded its registration and has consented to "agree with the terms of Facebook after reading our Data Policy, including our Use of Cookies." Obviously almost no one will actually read the twenty-two pages of information that, besides detailing a few rules to be obeyed in the social media's use, make it quite clear what the goal of Facebook's is: "we want our ads to be as interesting and relevant as all the other information found in our Services.[18]" In other words, to keep ourselves entertained not only by the services offered by the platform, but also by the gratuitous consumption of advertisements. Jean Baudrillard (1993) used to say that advertisement is the most democratic product there is: it is the only one that, escaping the logic of selling, is freely "offered" to everybody—and people do not have to pay a dime to be able to enjoy them. However, in the case of Facebook, not even the service has to be financially paid: one only has to have a cell phone connected to the Internet to access it.

However, "everything in life has a price": in exchange for the "power to share",[19] the information that one likes, posts, adds, replies, announces,

[17] Data available at: https://www.tracto.com.br/quantas-pessoas-tem-acesso-a-internet-no-mundo/, accessed on 07/27/2020.

[18] Information available at: https://www.facebook.com/policy.php, accessed on 07/27/2020.

[19] Amidst this scenario where the "new capitalism metabolizes vital forces" (Sibilia, 2012, p. 33), transforming them into capitalizable mediums, we can see the emergence of an immaterial sharing (Mozzini-Alister, 2019) that is directly related to the transformations arising from an immaterial work (Lazzarato & Negri, 2001). In it, one does not invest only in production, but also in "the whole configuration of the 'reproduction-consumption cycle:' the

comments, plays, and does on the platform is used as an instrument to enhance "our advertisement and mediation systems".[20] Systems that feed a marketing chain that aims not only to "show you relevant ads inside and outside our Services",[21] but also wants to sell the collected information to potential advertisers. Thus, through the constant subtilization of advertising strategies, the "new techniques sustain the repression: the consumer interiorizes, through the very movement of consumption, the social stance and its norms" (Baudrillard, 1993, p. 185). And interiorizes it so well that the consumer also reproduces it[22]: for the majority of users, it is old news that Facebook sells their digital traces to marketing companies and profit from that commerce. But, just like me, they simply don't care. Between keeping one's privacy and becoming prevented from "being connected to the world," the option is not only to finish creating our profile, thus forgetting about the sordid details of such policy but also to, in exchange for the possibility of "finding friends faster," allow the App to search for the information stored on our smartphones contacts.

There you go. The App lets us know it is making the "login"—a combination of the computer terms "log" (register) and "in" (inside), a

immaterial work is not reproduced (and does not reproduce society) in the form of exploration, but in the way subjectivity is reproduced," point out Lazzarato and Negri (2001, p. 30). This happens because if, at a certain moment, the act of sharing was limited to presential exchanges of material, palpable, artifacts, today it is diluted and gains a new territory of practices emerging not only from the possibility to share in a regime where presence is not necessarily physical, but also through the sharing of elements that cannot be objectified as external to the subject. One shares their life and its energetic mechanisms as a good, and what could be more abstract? As a result of this constant capture by advertisement of the modes of communication, the outlines and the implications of the mechanisms of immaterial sharing become even more visible. That is the case because, as proposed by Lazzarato and Negri (2001), "advertisement is used not only to spread information about the market, but also to constitute it. It enters in an 'interactive' relationship with the consumers, turning not only to their necessities, but above all to their desires" (pp. 62–63).

[20] Information available at: https://www.facebook.com/policy.php, accessed on 07/27/2020.

[21] Information available at: https://www.facebook.com/policy.php, accessed on 07/27/2020.

[22] It is through these virtual spaces of circulation and sharing of the self that, beyond a mere communicative role, imperative advertisement strategies are also developed and used to invigorate the processes of immaterial work and consumption. This happens because, as pointed out by Antoun and Malini (2011), "the act of sharing must be concomitantly controlled and disseminated" (p. 9) because "when managing the act of sharing, capitalism is not concerned about controlling production, but about controlling the circulation, thus managing the time for the socialization of knowledge, of wisdom, of creation" (idem).

neologism for every time somebody wants to access a virtual account. By sufficiently programming the subtilization of the methods that incite us to desire, the process of transforming the circle of the "I" into a dot on the Great Us/Knot of Facebook is almost complete. Almost. Before that, it is necessary to validate the account and to verify if the phone number and the e-mail address that the user provided are not fake. Immediately, the user will receive a numeric code via e-mail or text message to be typed into the space designated for the insertion of a "confirmation code." Once this non-transferable step is confirmed, another one follows, this time to "add friends." This stage's duration will greatly vary depending on who is going through it: one may skip it, take five minutes or spend five, ten, fifteen hours asking for the friendship of "people you might know"—as the App trackers keep indicating. While one waits for the invitation to be accepted in order to compose the network of dots that integrate the profile's affections, the last and most important dialogue tab between App and person opens up: "Add a profile picture so that your friends can find you." One can also skip this part and use an anonymous icon as your profile picture. However, a faceless profile is a "weird" profile: distant, it neither connects to itself nor to the other. Then, it is time to give the profile a face. A face that not only portrays the way in which one wants to be seen but, mainly, that captures and brings forth the soul hidden inside the circle of the "self" in a way that its double can be recognized as a faithful replica of the best possible version that one has to offer.

The relationship between soul and image is nothing new. Since long, ancient populations—like the Native tribes in the United States and Latin America, and the Aborigines from Australia and New Zealand—refused to be photographed because they believed that such gesture would steal and imprison their souls in the image. That is the irony of the human condition: it is so close that it can only see itself through the eyes of the other. It can see feet, arms, legs, torso, but the act of waking up and opening one's eyes can never reveal its own face, the self-portrait. Would that be the reason why a charmed Narcissus let himself be swallowed by the water mirror that reflected the outline of his face, which he had, at that point, never seen? Neglecting the advice from the most remote ancestors of humanity, we prefer to believe that, instead of stealing the soul, it is by photographing it that we can unveil it. Narcissus not only became a martyr but also reinforced, in the trauma of death through the image, the indefatigable project of technically building devices that put the eyes outside the skull without the risk of drowning. In this urge to see ourselves as we are, we invented mirrors, photographic, cinematographic, video, and surveillance cameras … Cameras that, by the way, are inside telephones today, which in turn were converted into devices with mobile, sensitive, and connected planes capable of granting electronic eyes to the tip of our fingers. If in the past we were afraid of touching images,[23] at last we are free from such complex: today, safe and sound, we feast on selfies.

And it is so much so that the App itself invites us to execute two possibilities to give life to the profile's image: "Take a picture" or "Choose from the Gallery." Without fear I touch the screen and I automatically choose the first option: I point the front camera's eyes toward my face, I click on a virtual button and I obtain a frozen image of the instant as a photographic record of myself. I could have chosen to take a picture unrelated to my face, a landscape picture, for instance, turning the camera eyes away from my face. Different from mine, the device's convertible eyes can see both what I see and also my own face, which I cannot reach. But I prefer to show myself as I really am. I then extend my arms, grab the device and strike a pose for the camera's accurate shot. I take one, two, three, as many photos as necessary. I search for the most beautiful angle of my features. I also create an atmosphere, a climate, an air of mystery thanks to the choreography of tiny movements of eyebrow and chin and, mainly, by the expression my look externalizes. After all, there is an old saying that goes: "the eyes are the windows to the soul." Nevertheless, it seems that the windows to my soul are not always open: not satisfied with my tired face, I give up from using a selfie as my profile picture and I move toward the second option, "Choose from the Gallery."

Now, my fingertips go to the device's visual memory stored in the icon Photos/Gallery, a kind of photographic album no longer limited to the usual binding process and in which I can quickly select, send, and upload a profile image. Immediately, a series of linearly organized images bring back memories of places, trips, unusual situations, foods, encounters, and social events that I have experienced. And a lot of other selfies as well. I choose not any given picture, but one that was taken with sufficient care to attract the attention of my peers. After said picture is found, the only one unique enough to be an image and likeness of what I truly am, creator and creature, person and profile, reach a silent agreement: a disintegration

[23] According to Benjamin (1985), in the first photographs the "sharpness of these visages was scary, and one had the impression that the small human faces present in the image were capable of seeing us" (p. 95).

pact (Flusser, 2008) in exchange for the possibility of connection. I give up from simply occupying the external space that contains me as well as the space occupied by my body in order to become fragmented in the digital dust and incarnate in a profile exhibited on the device's smooth skin. I no longer exist only at my house, on the streets, on the beach, on the field, at the office. And I am also not encapsulated by arms, legs, head, torso, voice, vision, sense of smell, touch, and hearing. I let it go from the heavy solidity that gives volume to the biological substratum to lightly duplicate myself as information, data, pixels capable of becoming computed dots (Flusser, 2008): I accept becoming a mosaic not only of carbon atoms, but also, of digits, of electrical impulses translated into a binary code comprised of bits of "0" and "1".

Without any delays, the App answers me with a new dialog box and shows a spinning blue circle—the "spinning wait cursor"—with the phrase "Loading…". I prepare myself to exist in this other physiology capable of surviving outside the body, in which the outlines of my face are converted into "pixels." It is interesting to notice that, even though the pixel-dot is, *par excellence*, the digital image's unit of measure, it is not unitary, fixed, determined, and incapable of changing. The pixel is, intrinsically, a "variable" dot. In this case, variable to the human touch on the screen that sets off a numeric variable resulting from the mathematical combination of three basic colors: red, green, and blue. Together they generate more than 16 million possibilities of colors to be arranged in lines and columns. It would seem weird to expect that, by touching the skin on the face or its self-portrait printed on a photographic paper, pores and grains all of a sudden would go through a mutation. Today, however, we internalize the visual plasticity as a communicational habit due to the naturalization of the gesture of exchanging pixels by touching images. Such malleability of the digital substratum is the secret ingredient that incites us to the fulfillment of the grinding ritual of one's body: by accepting to be photographed by the device's eyes and to inhabit it as a digit, one is also adhering to the premise of "the spatiality of erosion" (Vaz, 2010, p. 200) as a possible space for the "self." No longer individualized, circular, unitary, and closed in itself, a divided fraction of myself becomes fragmented in order to incarnate in a double materialized as a profile page on the App's interface (Fig. 2.4).

At last, a profile. The simple process of bipartition is now concluded: I also become a singular dot. More than two billion people and I currently share a profile and have agreed to Facebook's terms in order to transform

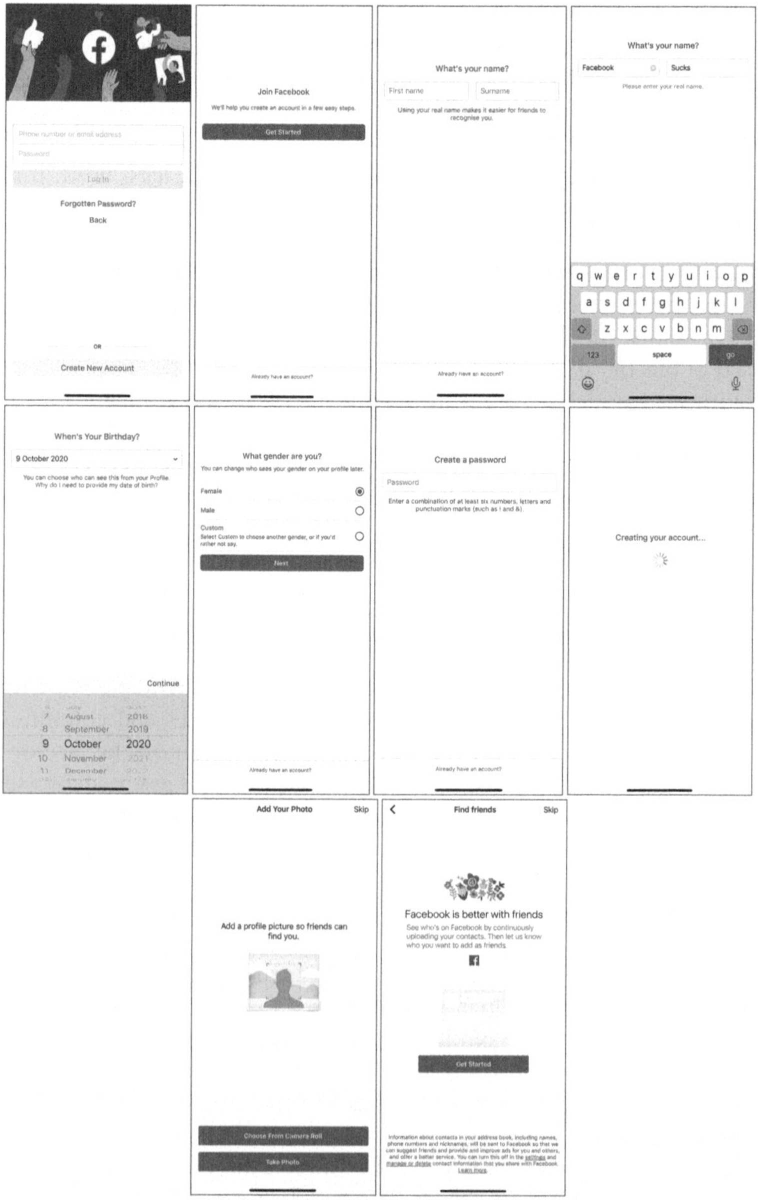

Fig. 2.4 Duplication process of the self as a profile, from Facebook App

our living into a better life because they are now connected to the fumbling of mutant dots. The surroundings that used to limit me to the geometry of the architectural space are now expanded to the space of a topological architecture[24] of the Great Us/Knot. "Topological," that is, made of a "series of dots connected to other dots" (Vaz, 2010, p. 201). Amidst this "entanglement of touching arrows" (idem), I am just another dot, but never a lonely dot. The web in which I enter is not any given web: it is not a fishing net, a sleeping net, a sewing net. I reach my very own core through a "social network." A network that allows me to see myself as I am by instantaneously connecting me to the remote gaze of others that are also seeing themselves through me. Thus, I externalize my soul, my "biography," my life's journey and incarnate in the smoothness of a smart skin by using the information of a personal page that is also my identification passport: I update preferences, work places, living spaces, favorite quotes, cover pictures, photo albums, and relationship status. To publicly announce the status of any new relationship on Facebook is a crucial social pre-requisite for said relationship to be considered "official." And if in the past my perception of reality encompassed "everything that we would bump into during our path to death" (Flusser, 1985, p. 5), now such perception is dilated. Immersed in the outlines of a virtual space that is mathematically engendered (Flusser, 2015), I no longer hit, trip, trespass, or bump into "things" or "persons" that physically knock down the roundness of my body. As an "information node inserted into a network whose wires transport data from 'self' to 'self'" (Flusser, 2008, p. 95), I only desire to enjoy the fact of having moved my soul "towards the particles that make the world" (Flusser, 2015, p. 96) while externalizing all its possibilities in the interaction provoked by the question: "what's on your mind?"

2.3 Screens and Waves

We are on the verge ofthe electromagnetism era. (Vilém Flusser)

[24]Whereas in the Euclidian geometric plane the structures are equivalent due to the isometric transformation of movements that conserves measures like angles, area, and volume, in the topologic field everything is deformed, except for the intersection between dots (Marcolli, 1974). To double in size, to shrink, to extend, or to bend, do not make a difference as long as one keeps the quantity of intersections between knots stable.

Two numbers in red, one right above the image of a bell and the other on the icon of two human figures, catch my attention and provoke a feeling of urgency: I received some notifications and some friendship requests were accepted. After all the stages required by the App were completed, I was able to transform the tangential line into a feedback channel between person and profile and duplicate the spatial dimensions of experience that allow us to enter the mysterious realms of thinking. From the disconnected circle that would only rotate around its own axis, conditioning the sharing of its ruminations to the scope of the physical presence, one has in its double a dot capable of making the last and final connection: to spread the individual mind's thoughts through a profile connected to the other two billion of "Us/Knots" tangled in the Great Mind of Facebook. This happens because, after the shock by the touch that turned the device into an animated body, a body possessing a soul thanks to the acquisition of a personal page, the profile dot also wants to have a mind. After all, what is the meaning of an empty profile, one that does not think about updates or does not expose its doubts and longings? If a device without being touched is a dead body and a profile without a picture is a soulless profile, a soul without a thought, without a timeline, is also a dysfunctional page: it does not make sense in existing. It becomes vital, then, the necessity of fulfilling this mental emptiness through the electrocution of fingers on the interactive structure offered by the App to connect the brain waves to the wireless waves that simultaneously allow for the irradiation of every update sent or received by the profile dot.

However, before entering this territory in further detail, how does the process that grants a mind to the profile actually work? Since it does not have an autonomous mind, capable of thinking for itself, the profile page urges: it urges to reverberate the thoughts of its creator, the only one who has the secret password to access it. Flusser (n.d.) would say that all mental processes, whether they be "perceptions, imaginations, desires, and decisions are chemical and electromagnetic processes reducible to quantum leaps of particles in the gaps between the human brain's neural synapses" (p. 3). And even though there are people who would say that the individual thought is nothing more than an anonymous whispering of historically conditioned voices,[25] in the end, it is inside the boundaries of the

[25] By considering the notion of discontinuity as method and effect, the genealogic conception of history proposed by Michel Foucault (2010) juxtaposes, to the authorial human scream, the anonymous whisper. A similar view can be found in Walter Benjamin (1994):

"I's" circle that such process takes place. We may regurgitate in our mind the most horrendous, visionary, or even the most insane possibilities, but no one will know what we are truly thinking unless we express the content of our thoughts through the use of words, gestures, facial expressions, stares, actions, and corporeal movements. As the Brazilian wisdom would say: "we think with our buttons".[26] That is why a part of the daily toil in dealing with humans is necessarily traversed by the difficult process of thinking and acting, imagining and expressing, deciding and exteriorizing our decisions. And because there is no expressed thought that does not have a physical action as its companion, in order to provide what the profile so badly wants, an action becomes necessary: to move the fingertips on the crystal screen where an insistent question shines like a mantra: "Camila, what's on your mind?".

Following the minimal gesture of typing a response to such question, my mental processes are instantaneously linked to the profile's thoughts. However, in this banal procedure there is a profound change in the very nature of the relationship between action and thought: if for the circle of the "I," to think and to act are two distinct instances, to the profile dot, the field of action is the thought itself. Dissolving such separation by the combination between thinking and acting as the only practical possibility, it is up to the profile to operate thoughts: "I post, therefore I am." Ultimately, the Cartesian philosophical maxim[27] is reinvented and the

with the proposition to tear up the veil of appearances, continuities, and essences in favor of the disruption of the hazardous and the fragmented, history emerges as an intensive and non-linear temporality, in which creation forces emerge as forces of transformation and rupture.

[26] In Portuguese, the expression: "*Pensando com meus botões*" (in a literal translation: "Thinking with my buttons") means thinking to oneself. It would be akin to introspection or musing.

[27] The moment Descartes (1979) publishes *Discourse on the Method* demarcates, not as an origin, but as a historical rupture, the emergence of the methodological question as a problematic issue for the so-called Humanities. Unhappy with the Humanities' lack of rational ground and with the overlooking of the use of mathematical knowledge in our daily life problems, he pursued the truth of the concepts via physical and mathematical demonstrations—considered by him as the only indubitable ones. Updating the Pythagorean ideal of submitting the universe to numbers, Descartes posited as methodological precept that only what is evident can be considered true. In this sense, he wanted to, by whatever means necessary, overcome any and all uncertainties, thus proposing the hyperbolic doubt as the method to arrive at the clarity of reason. Facing the hypothesis that we are being deceived by an *evil genius*, who would guide us toward making mistakes and would prevent us from seeing the objectivity of the material world, Descartes proposed the existence of a *benevolent God* who,

individual cogito that ruminates its doubts only with itself, becomes divided: by the clear and distinct substratum of digits 0 and 1, the double of myself becomes a being whose act is the immediate exteriorization of a thought performed to, and taking into account, the other. Other that is no longer limited to the physical spaces of interaction where the circle of the "I" wanders. The profile encompasses friendship networks remotely linked that expose the "I's" double to the mercy of other two billion dots that immediately react to the profile's socially shared "thinking-actions." Interwoven in the virtual space of network dots, that which I call the "'self' becomes a word that other people call 'you'. I only exist because of the others, and the others only exist because of me" (Flusser, 2015, p. 105). In this being for the other,[28] the reality parameters amplify the validation scope of what one is: what I keep to myself and dare to exteriorize through my double is worth being thought of?

Thus, by the physical gesture of gliding our fingers on the screen and placing what we think as a profile action in the form of texts, hyperlinks, images, pictures, videos, and live transmissions, three standard modes of potential reactions are configured: (a) the activation of the plug-ins for social interaction commonly known as "emojis"; (b) comments; and (c) shares.[29] Emojis are tiny images or digital icons, contemporary hieroglyphs

being infinitely good, would ensure that the image people have of the external world is not simply a creation of their minds. If God exists as being the guarantee for objectivity, the certainty that the material world is real is achieved. Thus, our thought offers to our spirit, through the methodical doubt, a series of rules to achieve true knowledge since it is the sensitive knowledge that causes the mistakes and errors, and truth is achieved by the intellect. Via a deductive process, it is the cogito that ensures "the existence of the 'thinking thing:' the soul" (Regis & Messias, 2012, p. 34).

[28] In this sense, complementing Flusser, Muniz Sodré (2007) points out that "the question about the act of communicating as creation of common emerges, on the other hand, from the relationship or the affiliation implicated by this 'with,' (*com*) that demarcates the division of a *munus,* a task or a gift originally performed or offered by one individual to another. To communicate is the action of always, infinitely, seek the common in the community, not as a being (for example, a group or a gathering of subjects), but as a link, therefore, as a constitutive nothing, since such link does not have any material or constitutional substance, it is a pure breach in the language. The subject who is communicating is the same being as "in between," hence, an interiority destined to an exteriority, the other" (Sodré, 2007, p. 21).

[29] It is interesting to point out how the notion of sharing assumed, since 2004, a new meaning with the advent of Facebook. From the Latin "*compartiri,*" the word is derived from the linking of the prefix "*com*" (with) with the words "*partiri*" (to divide, to distribute, to share) and "*pars*" (a part of, what is common to everybody). If for a long time the composition of the word "sharing" translated a wide array of practices related to the act of shar-

used to express an idea or emotion in electronic communications that, in the Facebook case, are six in total: "Like," "Love," "Haha," "Wow," "Sad," "Angry" and their respective drawings of a thumb pointing up, a pink heart, a smiley face, a face with an open mouth, a crying face, and an angry face. For their part, comments are free reactions or open answers directed at certain posts that can take place in the form of text messages, images, photos, hyperlinks, or videos. Finally, shares are dubious reactions: at the same time that they can act upon an already established thought, they can also be considered reactions used to hold the person who originally posted accountable. From the combination of these three types of potential reactions we have the selection of programmed possibilities of social interactions for the profile that kick-start the instantaneous game of actions and reactions, of posts in search for likes, commentaries, and/or shares.

But what would make possible for actions and reactions to flare up like that, so "instantaneously?" Such question can only be answered in one way: by steps. Before a phenomenon that, for being so fast, loses its condition of process, we can only slow it down to the point where it becomes once again capable of producing the feeling of awe. In slow motion, we will start then by the dimension of action, that is: by the act of posting and mentally linking person and profile. We do not know exactly where our mind is located, but we know that it exists because it thinks. By the way, it thinks so obsessively that it cannot stop: at all times we are devising a plan, remembering an episode of our past, projecting a desire into the future,

ing something in a community, to share, today, has a new status and another semantic structure since it is associated not only with the materiality of groups but also with the crowd of millions of Internet users that, in an ethereal way, constitute and are constituted by, and through, digital networks. However, we cannot single-out Facebook as "the" catalyst of the current wave of sharing we are witnessing: less than being the cause, the act of sharing led by Facebook arises from the socio-technical transformations that occurred during the passage form the twentieth to the twenty-first centuries, consolidating the so-called Web 2.0. Even if it sounds sequential, the term Web 2.0 is related not to a new version or an update of the Internet, but to qualitative changes in both the way software programmers as well as Internet users used to employ digital platforms. It is within this context that emerges a different browsing architecture based on models such as blogs, social networks, online encyclopedias, websites that share videos, Apps, thus ending that rigid separation between producer and consumer of content, for the focus would no longer be on the isolated transmission, but on the participation and the collaboration. To Zanetti (2011), "it is only after the emergence of the so-called Web 2.0 and the advent of social networks that a discourse putting in evidence the idea of sharing becomes sanctioned, actually legitimizing such practice" (p. 65).

trying to solve daily problems, searching for a creative solution to our doubts, philosophizing, asking ourselves where we came from and where we are going. Crammed with mental projections, the chaos of images that are taking place is nothing more than chemical processes intertwined with wave frequencies that vibrate throughout the nervous system's neural synapses. And to avoid the risk of imploding in our own minds and falling into the abyss of madness—or as Deleuze (2007) would say, the loss of strata—we find a way to unload the unbearable weight of our own thoughts through the use of mnemonic supports. Sand, caves, papyrus, canvas, notebooks ... Today, one of the ways to accomplish such process is to grab a smartphone, open the Facebook App, electrocute the fingers on the obedient imagetic keyboard and externalize what we are thinking in the form of pixels or computed dots (Flusser, 2008) immediately visible to a double comprised of dots inside a space of dots algorithmically structured.

Done! Just like magic, the waves of the mind are exorcized as waves that circulate inside the device's electronic circuits, which are in turn converted into bits, that will become waves once again, but this time, radiofrequency waves[30] to be irradiated by the device's inner antennas that will navigate modems, antenna towers, underwater cables and/or satellites until they can communicate to the external antennas of Facebook's servers, which will ultimately receive the data, update it as a profile's "thinking-action" and send it all the way back until it becomes pixels that will flicker on the skin of my smartphone's App. In an imperceptible "minimal before-after interval" (Hansen, 2009, p. 310), each time we inject a thought into the profile, three distinct circuits are synchronized into a single electromagnetic circulation circuit: the neural vibrations inside the body, the electrical vibrations inside the device's body, and the radiofrequency vibrations irradiated via antennas, submarine cables, and satellites. A big wave created by the nervous symbiosis of wave systems that are, at first, separated. It is by means of this electromagnetic tsunami that we externalize as far as possible the inner waves of the mind. First to the smart device that soon will take care of irradiating them from the bottom of the ocean to the

[30] In general, devices that possess an antenna will communicate to one another through the irradiation of radiofrequency waves. Wide as buildings or mountains, slow and with little energy when compared to microwave, terahertz, infrared, ultraviolet, X-rays, and gamma rays frequencies, radio waves are considered the least potent waves in the non-visible spectrum. Even so, they represent the most profitable and economically attractive segment of waves related to phone carriers, and are internationally regulated by the International Telecommunication Union (ITU), related to the UN.

stratosphere's most remote layers only for them to arrive, safe and sound, at their two final destinations: the timeline and the newsfeed.

How many things happen in what we do not see? At a speed so high that escapes the field of the humanly visible, capable of only seeing frequencies between 400 and 700 nanometers,[31] our thoughts are indexed not only to the archive shown on the profile's page but also to the collective display of the newsfeed. As a communal showcase for the announcement of posts, in it the tragic dispersion of "each profile by itself" is avoided by the gathering of minds programmed to live together. "Post," a neologism from the verb "to post" that comes from the Latin "*ponere*," a reference to the act of placing, putting something somewhere. And based on a programmed architecture of posts placed vertically one after the other, to give continuity to the visualization of what comes right below or above depends on the execution of a second physical action: the zigzagging of fingers that go up and down on the screen. Known as "scrolling," which means "to displace," "to move," "to go through," it is only via this gesture that two possibilities open up: we not only see ourselves by virtue of the foreign gaze of retinas, cameras, and screens, but we also see ourselves better as a result of the comparison due to the inexorable spinning of this papyrus of digital thoughts. Finding oneself has now become a synonym of losing oneself in images, videos, texts, and pictures. And taking the wisdom that "to exist is to forget about oneself" (Flusser, 2015, p. 105) to its last consequences, we simply defocus: we defocus from what surrounds us and we defocus from the body in which we are contained in order to, just like hamsters, keep playing on this carrousel of vertical images (Fig. 2.5).

Immersed in the profile's mental reality and its holographic substance palpable only through the shock of fingers on the screen, we feel ourselves moving even though we are standing still: even when the body is not in movement, the human mind spins in line with the whirling of the thoughts of others. Subtly, we are no longer orbiting the work we have to do, the decisions we have to make, the food we want to eat, the conversations we need to have, or how tired we are feeling. Now we orbit around the profile's dot as a new axis capable of going through a conglomerate of minds on the newsfeed. Unbeknownst to us, all of a sudden the rotation of the circle of the "I" around itself has stopped and is converted into a translation movement around the thinking-actions of its double. And on such

[31] A nanometer is equivalent to a billionth of a meter, that is, a meter divided by a billion.

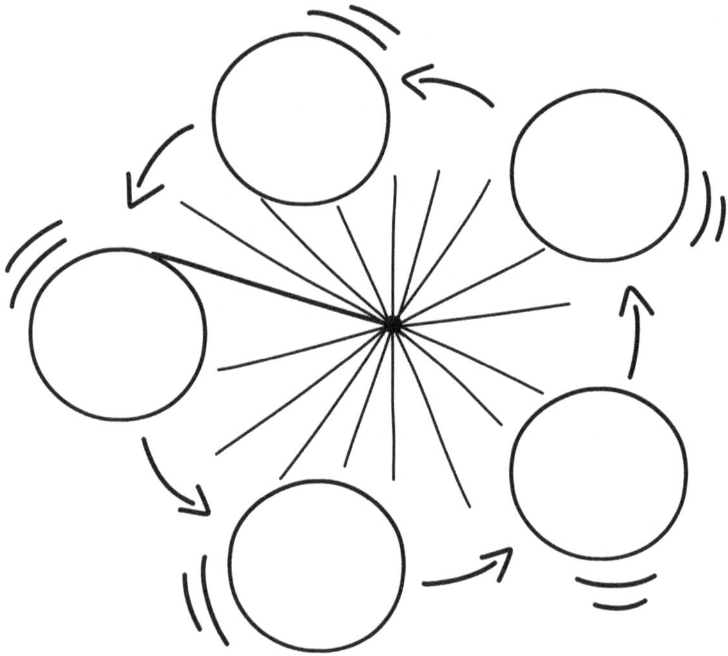

Fig. 2.5 The translational movement of the "I" around the profile, by the author

actions, we can simply pause to quickly distribute reactions: I like the dog video posted by a friend from São Paulo, I love the picture of a trip to Canada that my friend from Berlin went on, I comment on a post of my doctorate colleague from Rio de Janeiro, I get sad with the death of my friend's father from Rio Grande do Sul, I like the advertisement of a new Japanese dermatology product, I get angry with the link to an environmental denounce posted by a friend from Valencia, I tag my Australian partner in the vegetarian recipe posted by my aunt in Passo Fundo, I say "Wow" to the selfie showing the new look of an acquaintance from Melbourne, I comment on the tagging of a friendship's birthday post made by a friend who lives in Teresina, I like and comment on the commentaries people wrote on something I recently posted, I love the video of Angolan children dancing published by my friend living in Recife, I share the ideals published by a friend from Córdoba, I send a message offering help to a friend I met during my Master's studies in Porto Alegre,

I like a sponsored page about astronomy videos located in the Unites States, I comment on the timeline of a childhood friend from João Pessoa, I love the poem written by a friend who had just been to Colombia, I share a memory from exactly five years ago sent by the Facebook archives, I send a longing happy birthday to Diana's profile, my roommate who slipped and tragically died as she was taking a never-published selfie on the cliffs of the Greek island of Zakynthos.[32]

With every reaction, an affective trace is materialized like footprints on the pixels' digital sand. I leave a vestige of myself through the traces of my double, which reaches and touches those who are no longer within the reach of my legs, hands, ears, nose, tongue, touch, sex, or vision. Although I am not physically in any of these places, I exist in all of them. By spectralizing myself in the "immaterial materiality" (Michelin, 2008, p. 170) of the trine circuit of electromagnetic waves, I multiply the dots of the "Self" in the dots of the "Others" as a result of the desire to connect body, soul, and mind to every other device, profile page, and timeline of the network. Supported by the infrastructure of around 1714 satellites,[33] 375 underwater cables[34] and countless antennas, this web of interceptors allows for the immediate unveiling of the interplay between action and reaction that expands my scope of presence toward the outside. It is as if having the Facebook App on my smartphone "would create in me a series of cables scattered all around the world that would work as an extension of my body"—said my friend Giulia when we met face-to-face in Valencia around April of 2016. And through the sharing of this universal and interactive code organized around six emojis and two tools for commenting and sharing, the energetic symbiosis between body and device is converted into an affective symbiosis that expands the body's sensorial radius across the body of the device. Now linked to the smart body's ingenious sensors, the

[32] Besides being beautiful and full of life, Diana was also sweet, highly intelligent, wise, and was always smiling. Her dark brown eyes and her big Amazon hands will never be forgotten. Information about the episode that became international news can be found at: https://www.thetimes.co.uk/article/american-tourist-diannarosa-alduncin-delgado-dies-taking-selfie-on-cliff-in-greek-island-of-zakynthos-hxxjrqmg6 and the Instagram hashtag #livelikediana, accessed on 07/27/2020.

[33] A quick observation using the Satellite Viewer website makes it possible to interact in real time with each one of the 16,000 satellites, among which 3000 are still active, that orbit the planet: http://apps.agi.com/SatelliteViewer/, accessed on 07/27/2020.

[34] A complete and interactive map of the underwater cables that exist today can be found at: https://www.submarinecablemap.com/#/, accessed on 07/27/2020.

biological substratum thus acquires a new array of senses: what is felt as "body" is also activated by the intimate shock between the carbon and the crystal skins in the App's graphic interface, which communicates with antennas and irradiates affective tentacles across the four corners of the globe (Fig. 2.6).

For the skeptics, one has only to remember a case, any case, in which drastic resources of reaction became necessary: unfriending a friend or blocking a profile. Even though these are reactions addressed specifically at a profile, their affective consequences directly impact the existential territory of those who are reacting and also those who are suffering the effects of such reactions. We exclude or block in order to defend ourselves from something that is considered an invasion of personal limits: limits that now integrate the profile's topological space as a living layer of our vital space. By virtue of the bond created between the person and its double, the body's sensorial perimeter is amplified in favor of the possibility to

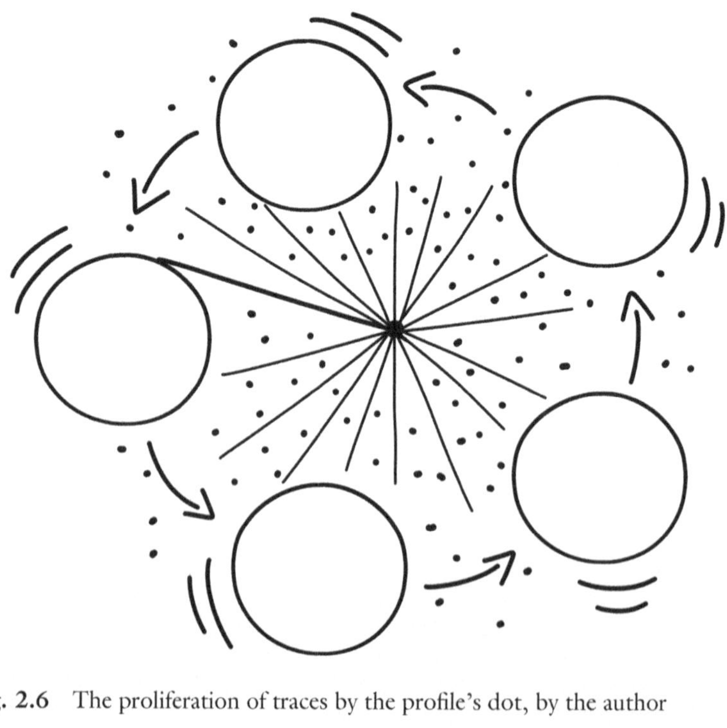

Fig. 2.6 The proliferation of traces by the profile's dot, by the author

remotely feel. To feel even when one is geographically distant. It is within this other form of presence, of this "feeling of being the embodiment of something" (Gumbrecht, 2010, p. 167) that the ambiguous experience of a distant proximity takes place: proximity between people through the distance of profile dots that touch one another by the touching of fingers on Apps and screens of smartphones connected to a series of antennas, cables, and satellites. And because it is a presence untethered from the physical experience of space, such proximity is only measurable by the intensity of time: the more simultaneous the interplay between action and reaction, the more strongly we feel the presentification of a body that is not physically closer to us. In this consists the indefatigable search for the touch of the "Other's" gaze on social media: it is through the immediate exchange that the "I" experiences the reciprocal acceptance of the alterity of what she/he is in real time: being.

However, the expectation of immediate reciprocity by the utmost number of likes, loves, laughs, surprises, anger, sadness, comments, and shares can be frustrated. In this scenario, impatience, aggressiveness, rejection, failure, spite, confusion, and lack of confidence are a few of the possible affections that will compose the other side of the coin to be ruminated when one savors the bitter taste of invisibility. From this point on, two main roads are possible: to invest even more fiercely in the process of self-comparison and to keep checking for notifications, spinning the wheel of images on the newsfeed, distributing reactions, and incessantly posting new *thinking-actions*. Or one could simply quit the network, disconnecting one's body from the device. There are cases in which the renunciation is absolute and one can deactivate or even delete its social media account. Nevertheless, the majority of circles take another path. Back to the volumetric reality of space and to the processual slowness of time, soon they bump into a doubt: "what have I lost while I was disconnected from my dot?" Tormented by the lack of the element that expands it toward the outside of its own body, the circle of the "I" once again grabs the smartphone, logs into the App, and keep chasing its goal: the integral communion with its double. However, such ambition can only be achieved by means of a double strategy: to make the circle of the "I" as big as the world and, at the same time, to fragment it in a way that it becomes just another tiny dot connected to all the other profile dots.

Driven by the desire to constitute an extra-corporeal body, the execution of these two simultaneous procedures begins: to incorporate the device's body in order to incarnate its double on the screen. Let us begin with the incorporation process: a Latin word composed by the combination of the prefix "*in*" (toward the interior) and the root word "*corpus*" (body) that refers to every process of introducing something in order to create another body, indexed to what has been added to it. Thereby, it is by the aggregation to the device's body that the combination between person and profile accomplishes the acquisition of two new senses concerning the biological substratum. First, to equip the fingertips with eyes capable of seeing themselves through their own touch and, after that, to give arms to these mutant hands that are transformed into electromagnetically stretching arms capable of touching everything that they have seen related to others who are geographically distant. With this enlarged body, we now have the sharp feeling of being able to embrace the world through the wireless extension of arms that runs in parallel to other process: the incarnation of this spectralized body in the glowing skin of the crystal screen. From the Latin "*incarnare*," the term refers to any process of representing someone in a material form, personifying an abstract concept. In this sense, it is the "I's" double—the mental substance archived on the profile page—that becomes flesh by occupying and manifesting itself into the device's body. As a dot, it moves across the network irradiating itself through all the devices' bodies capable of accessing Facebook. Always accessible, or perhaps not: if someone "hacks" the account and/or overtake its online identity, that may cause a certain headache to the creator of such creature.

Henceforth, the interplay between these two complementary inverses is consummated: without ever dying, the profile incarnates and reincarnates in any other body of device that decides to access it; and unwilling to move, the body attached to the body of the device launches its arms equipped with eyes on its fingers until it touches as many dots as it wants to see on the network. In this orgy of profiles that are mutually exciting one another through the indirect touch of each other's intimacy, doses, and more doses of dopamine[35]—a neurotransmitter that acts on the control of movement, on the memory, and on the feeling of pleasure—are liberated by the stimulation of the sympathetic nervous system. With the acceleration of one's heartbeat and the increase of the adrenaline levels, with the dilated bronchi, the motility of the large intestine diminished, the constriction of blood vessels and the dilated pupil, one reaches an orgasm.

[35] Several studies are increasingly relating the use of social networks to the circuits of hormonal liberation of dopamine in the body. More information at: https://www.epochtimes.com.br/psicologa-analisa-vicio-facebook-midias-sociais/, accessed on 07/27/2020.

A cerebral orgasm conflagrated in the much-desired "mission of making the world a more open and more connected place"[36] through the unveiling of the last and final of Facebook's ten principles: "One world": "Facebook's service should overcome geographical and national barriers and be available to everybody in the world".[37] Thus, all dots are connected to one single big dot. By turning the "I" not only into a circle, but also into a dot within the reach of any device on which I scatter myself via the "mosaic of tiny stones" (Flusser, 2008, p. 134) of the "telematic superbrain"[38] (idem): I multiply myself and become always accessible to those who are distant from my touch. By making the body not only a body comprised of limbs, torso, and head with its respective five senses, but also a body whose senses, amplified by the device's body, resize the diameter of the circle of the "I" to fit the circumference of planet Earth: I make myself the size of the world so that I can reach the ones I love and who are no longer by my side. In this planetary circle covered by connected dots, at last I am one with everything: one with the Great Mind of "Us/Knots" weaved by Facebook (Fig. 2.7).

***

There I was, at the beginning of 2014, unfocused and intoxicated as a result of feeling limitlessness. No longer inaccessible when eating, sleeping, reproducing, working, meditating, and studying. No longer accessible only to those whom I have physically crossed paths with during my daily life. Without needing to move arms and legs, I only have to perform bare minimum fingertip movements and keep my eyes always open to, effortlessly, extrovert myself as far as possible from the body. With the fingers ready to glide on the smartphone's smooth skin, I touch the other

[36] Information available at: https://www.facebook.com/policy.php, accessed on 07/27/2020.

[37] Information available at: https://www.facebook.com/principles.php, accessed on 07/27/2020.

[38] According to Flusser, this is the idea behind the so-called contemporary cultural revolution (Flusser, 2008). A revolution that unites two distinct but convergent trends: the computation of punctual elements on surfaces, under the umbrella of the term "informatics," and the transmission of punctual elements through the so-called telecommunication. Two complementary trends that, even though simultaneous, for a long time were considered isolated phenomena. Photography and telegraph, films and telephone: even though pictures could be telegraphed and films could be "telephonable," something was missing—the idea that both could be technically fused, amalgamated. And it was only after the emergence of the duo video and cable systems that the click finally happened: at last, dots and rays, information and communication were coupled through computers under the term "telematics."

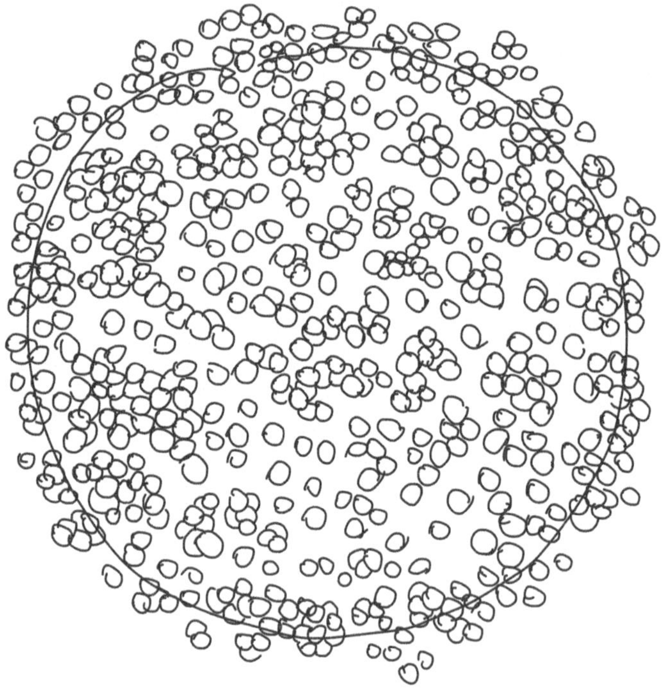

Fig. 2.7 The Great Mind of Facebook's "Knot/Us" and the union of the "I" with all the dots of the network, by the author

through the indirect touch between profile dots that share affections via posts, comments, shares, and the six emojis "Like," "Love," "Haha," "Wow," "Sad," and "Angry." Lost in order to be found in the electromagnetic communion between brain waves, device's waves, and radiofrequency waves irradiated by an infrastructure of antennas, underwater cables, and satellites until it arrives at the showcase of posts and profiles' pages. At the speed of light, anywhere and always present: connected, as connected as an "I" could possibly be. So connected that I ejected from the circle of the "I" a profile dot and created between them a path of mutual feedback that orbits around my double, multiplying its dots into a great planetary dot capable of embracing the world. The distances that kept separated what is far away and what is close by have disappeared. The circle that thought to be alone now scrolls the individual mind through the spinning of minds on the newsfeed. By embodying new senses via the body of the device and incarnating mind and soul through the apparatus, the interplay between actions and reactions gives life to the simultaneous cycle of collective validation of what can be thought. Electrocuting myself through the shock of fingers on the screen, the "I" is enlightened by the "Us/Knots." Perpetually reachable by the virtual presence of my double, the feeling of proximity emerging from the distant touch has increasingly separated the "I" from the actual human touch. A touch more and more distant from the body, the flesh, the family in João Pessoa, the friends in Porto Alegre, Rio de Janeiro, Brazil, Latin America, Europe, Australia: from any continent. An absent touch that would mitigate the longing for those who were far away through the touch on the screen, which ended up turning the habit of mediation, into an addiction.

References

Antoun, H., & Malini, F. (2011). Controle e Biolutas na Cibercultura: monitoramento, vazamento e anonimato na revolução democrática do compartilhamento. In *Anais do XX Encontro Anual da COMPÓS* (pp. 1–17). UFRGS.

Ascott, R. (2010). *Homo telematicus* no jardim da vida artificial. In A. Parente (Org.), *Tramas da rede: novas dimensões filosóficas, estéticas e políticas da comunicação* (pp. 241–247). Sulina.

Baudrillard, J. (1993). *O sistema dos objetos*. Perspectiva.

Benjamin, W. (1985). Pequena história da fotografia. In *Obras Escolhidas I*. Brasiliense.

Benjamin, W. (1994). Teses sobre o conceito de história. In *Walter Benjamin—Obras escolhidas*. Vol. 1. Magia e técnica, arte e política. Ensaios sobre literatura e história da cultura. Sérgio Paulo Rouanet (trad.). Brasiliense.

Bruno, F. (2010). Mapas de crime: vigilância distribuída e participação na cultura contemporânea. In F. Bruno, M. Kanashiro, & R. Firmino (Orgs.), *Vigilância e visibilidade: espaço tecnologia e identificação*. Sulina.

Bruno, F. (2012). Rastros digitais: o que eles se tornam quando vistos sob a perspectiva da teoria ator-rede? In *Anais XXI Encontro Anual da Compós, 12, 12, 14, 15 de junho* (pp. 1–18). UFJF.

Coccia, E. (2010). *A vida sensível*. Cervelin, Diego (trad.). Cultura e Barbárie.

Crary, J. (2010). A visão que se desprende: Manet e o observador atento no fim do século XIX. In L. Charney & R. Schwarz Vanessa (Orgs.), *O Cinema e a invenção da vida moderna* (pp. 81–114). Cosac Naify.

Deleuze, G. (2007). *Francis Bacon: lógica da sensação*. Roberto Machado (trad.). Jorge Zahar.

Descartes, R. (1979). *Discurso do método; Meditações; Objeções e respostas; As paixões da alma; Cartas; introdução de Gilles-Gaston Granger; prefácio e notas de Gerard Lebrun; tradução de J. Guinsburg e Bento Prado Júnior* (2nd ed.). Abril Cultural.

Flusser, V. (1985). *Filosofia da caixa preta: ensaios para uma filosofia da fotografia.* Annablume.

Flusser, V. (2008). *O Universo das Imagens Técnicas—Elogio da Superficialidade.* Annablume.

Flusser, V. (2015). *Comunicologia—Reflexões sobre o futuro.* Martins Fontes.

Flusser, V. (n.d.). *Luzes.* Texto inédito preservado no Flusser Archiv.

Foucault, M. (2010). *A arqueologia do Saber. 7.* ed. Luiz Felipe Baeta Neves (trans.) Rio de Janeiro: Forense Universitária.

Gumbrecht, H. U. (2010). *Produção de Presença*: O que o sentido não consegue transmitir. Tradução Ana Isabel Soares. Contraponto: Ed. PUC-Rio.

Hansen, M. (2009). Living (with) Technical Time: From Media Surrogacy to Distributed Cognition. *Theory, Culture and Society, 26*(2–3), 294–315.

Hennigen, I. (2011). *Publicidade do crédito: que práticas de consumo e endividamento são incitadas?* Alegre.

Kastrup, V. (2010). A rede: uma figura empírica da ontologia do presente. In A. Parente (Org.), *Tramas da rede: novas dimensões filosóficas, estéticas e políticas da comunicação* (pp. 80–90). Sulina.

Lazzarato, M., & Negri, A. (2001). *Trabalho imaterial—Formas de vida e produção de subjetividade. Tradução Mônica Jesus.* DP&A.

Lemos, A. (2012). Espaço, Mídia Locativa e Teoria Ator-Rede. In *Anais XXI Encontro Anual da Compós, 12, 12, 14, 15 de junho* (pp. 1–15). UFJF.

Marcolli, A. (1974). *Teoría del Campo—Curso de educación visual.* IBERDOS.

Marks, L. U. (2002). *Touch: Sensuous Theory and Multisensory Media.* University of Minnesota Press.

Michelin, S. (2008). O espaço-entre. In *Estéticas tecnológicas: novos modos de sentir* (pp. 169–174). Educ.

Mozzini-Alister, C. (2019). *Impressões de um corpo conectado: como a publicidade está nos incitando à conexão digital.* Appris.

Musso, P. (2010). A filosofia da rede. In A. Parente (Org.), *Tramas da rede: novas dimensões filosóficas, estéticas e políticas da comunicação* (pp. 17–38). Sulina.

Parente, A. (2010). Enredando o pensamento: redes de transformação. In A. Parente (Org.), *Tramas da rede: novas dimensões filosóficas, estéticas e políticas da comunicação* (pp. 91–110). Sulina.

Recuero, R. (2009). *Redes Sociais na Internet.* Sulina.

Regis, F., & Messias, J. (2012). Comunicação, tecnologia e cognição: rearticulando homem, mundo e pensamento. In F. Regis, A. Ortiz, L. C. Affonso, & R. Timponi (Orgs.), *Tecnologias de Comunicação e Cognição.* Sulina.

Sibilia, P. (2012). *Redes ou paredes—A escola em tempos de dispersão. Tradução Vera Ribeiro*. Contraponto.

Sodré, M. (2007). Sobre a *episteme* comunicacional. *Matrizes* (USP), São Paulo, v. 1, n. 1, out, pp. 15–26. Retrieved June 18, 2020, from http://200.144.189.42/ojs/index.php/MATRIZes/article/view/3987

Vaz, P. (2010). Esperança e excesso. In A. Parente (Org.), *Tramas da rede: novas dimensões filosóficas, estéticas e políticas da comunicação* (pp. 189–208). Sulina.

Zanetti, D. (2011). A cultura do compartilhamento e a reprodutibilidade dos conteúdos. Ciberlegenda, n. 25, Rio de Janeiro. Retrieved February 14, 2020, from http://www.uff.br/ciberlegenda/ojs/index.php/revista/article/view/496

CHAPTER 3

Meditation and the Process of Inner Expansion

July of 2014. I remember that my last meditation was horrible. Everything hurt: my legs, my back, my mind. My thoughts were tormenting me in those twenty minutes that seemed to last forever. As usual, at the end of every practice we did the "*Guru Puja*," that is, we sang a mantra in Sanskrit aimed at offering each and every thought that we had while sitting in lotus position during our meditation. On that day, I offered my meditation. And after that painful experience, it seemed that the prayer offered in my last *Guru Puja* was heard: without expecting, calculating, thinking or even controlling, my ***kúlakundálini***—from the Sanskrit "coiled serpent," or latent spiritual energy at the base of the spine—went up and, all of a sudden, everything became golden. Even though my eyes were closed, I could brightly see my surroundings shining in the most golden of lights. A light that was taking me to the depths of an indescribable force, the strongest one I have ever felt. In flames, I felt my body boundless, burning with love and merging with that intense glow.

Sitting in lotus position, my back was as stiff and straight as an iron bar, my body was shaking as if I was having an epileptic seizure and tears were streaming down my face. Tears of joy, from the most joyful happiness I have ever felt. The thirty minutes designated for the collective meditation had ended and I was still there, in ecstasy, sitting, laughing, and crying at the same time. And it was then that said happiness turned into sadness: wanting to placate my seemingly emotional distress—after all, from the outside, that scene seemed to be an uncontrollable existential crisis—some

C. Mozzini-Alister, *Does Social Media Have Limits?*,
https://doi.org/10.1007/978-3-030-74120-4_3

people came closer and started to hug me, to caress my hand: to touch me. With every touch, I felt my body coming back. With every touch, I stopped being light. I wanted to fend off those hands and say "no!", but I just could not talk or move. Slowly, the hands of the others were artificially waking me up from my ecstasy. The feeling was akin to an "interrupted intercourse." I resisted that for thirty more minutes until, at last, I opened my eyes with one single desire: to experience that again but, this time, without anybody around to touch me.

Touch. Where did my body go? Where did that pain that consumed my back went? Why did my voice disappear and why did my hands become numb? And what all those touching hands were taking from me? Six months ago, when the practice of meditation found me in January of 2014, I agreed to try it due to the despair caused by a deep depression. Having "reconnected" myself after a week fully immersed in a spiritual retreat, meditation stayed with me and, since then, I not only needed but I also truly wanted to meditate. However, after the golden experience, the act of meditating gained another dimension: I decided to take it seriously. I firmed my body in the discipline of *ásanas*, mantras, vegetarian diet, and two daily twenty-minute sessions of meditation. Every day, I would wake up 1 hour 30 minutes before any other scheduled activity to dedicate myself to my personal practice and dive for hours into books to study about the "liberation of the mind."

I was obsessed and in awe with that new universe: all of a sudden, I had become a "yoga person." And it seems that I was not alone in this trend: it is not uncommon to see, even among my close friends, more and more people completely changing their habits after turning to yoga in order to acquire a new perspective on the world. Confined in the past to the Eastern bubble, today meditation is a kind of "fever": electrodes are attached to the heads of monks with the goal of mapping their brain function during the meditative state; videos that teach how to practice meditation in under a minute[1] became viral on audiovisual platforms; yoga teachers are infiltrating the highest corporative echelons due to CEOs using meditation as a way to improve productivity; and the concept of

[1] To those who are interested, here is the link: https://www.youtube.com/watch?v=F6eFFCi12v8, accessed on 07/27/2020.

mindfulness,[2] besides being increasingly disseminated, is starting to integrate public health systems.

Are you feeling sad? Start meditating. Trouble focusing? Meditation will help you a lot. Want to relax? Meditate. Insomnia? Meditation. Just like magic, each and every ailment can be miraculously cured through the act of meditating. Nowadays, meditation became a new wave that curiously is being ridden in parallel with the explosion of sociability forms weaved via social media profiles: not coincidentally, Facebook daily asks me "What's on your mind, Camila?" The question about "what is on your mind?" is increasingly becoming an object of empirical investigation in the so-called Western world. But after all, what in blazes is the "mind"? What neuroscience is currently making palpable on "this side" of the planet through sophisticated brain mapping techniques, without any sort of fancy equipment, the "other side," already thoroughly and intuitively investigated thousands of years ago. Such investigation was undertaken by yogis who today are leaving the reclusion of the Himalayans to scatter themselves across the chaotic noises of the city life. But what is the relationship between the crescent desire to meditate and the modes of expansion of the biological substratum? And how is the desire for omnipresence manifested through the empirical process of meditation?

Based on such inquiries, we will begin the study of our second anthropotechnic method: meditation. Inspired by the personal experience of Tantra Yoga's practical philosophy proposed by P. R. Sarkar (or Ánandamúrti), we will investigate the desire for limitlessness no longer mediated by smartphone screens, but by the screens of the mind. As an approximation substratum for the divine touch, we will proceed a material archaeology of the mind through three stages: like an inverted mirror (Bernardo, 2011) of the mediation practices, in the first stage we will study the waves subtilization processes of the biological substratum that prepare the body for the practice of meditation; after that, we will describe the dotification process of other layers of the mind that are no longer visible to the human eye; and finally, we will thoroughly look into the procedures of shock of the dotified mind by mantras that aim to merge the individual mind with the Cosmic Mind. We hope that at this risky stage, we are able to escape unscathed in our endeavor concerning the modes of

[2] To know more, you can access the link: https://www.mindfulnessbrasil.com/, accessed on 07/27/2020.

approximation between body and light in the search for the enlightenment of the mind.

3.1 Mind and Waves

This universe is made of vibrations. (P. R. Sarkar)

If in the practices of mediation, to visibilize the body's undulating existence is the final stage to be scrutinized, in the meditation it is, straight away, the starting point: everything that exists, whether being visible or invisible, is nothing more than vibrations of different wave frequencies. With this vibrating body, we inhabit the molecular level and, with it, we constitute the outlines of a palpable, visible, and delimited form. A body through which we love, get angry, sad, and tired, we agonize, we move, we stop, we feel, we touch: we are affected. Nevertheless, this body covered by a skin that protects organs, tissues, and liquids through which we confidently feel to be ours cannot sum up what we are: there is something else, beyond vision, beyond touch, smell, taste, and hearing. Beyond the senses. Beyond the skin. Beyond that which one can manipulate with gestures. Beyond the usual perception of reality. Beyond books, treaties, writings, and documents. Beyond the physical existence of matter. Beyond the body. Beyond: something accessible only through the radical experience of empirically unfolding the five senses in order to penetrate the depths of the mind's realms.

But if what we are is not what we think to touch, and much less what we touch on a smartphone screen, what are we, after all? Why do we have to confuse what is already complicated? Why do we need to know who we are? Is it possible to grasp the concreteness of the "I" if it appears to us always so distant, fleeting, and fragile? At some point in the beginning of 2014, coincidently at the same time as I was starting my Ph.D. studies, questions like these became inescapable: I wanted to know once and for all "who am I." Drowning in the tears of a profound depression, more and more I wanted to dive deeper and deeper into the waters of the repugnant sea of "interiority." Repugnancy: academically trained to brush aside any approach that suggested the existence of "inner truths metaphysically anchored," I had only two options left: to live a life regulated by antidepressants, or to swallow my intellectual pride and go look for something unknown. I decided to take that risk and unbeknownst to me, by choosing the second alternative, I embraced the second vector of the desire for expansion and deconstruction of the biological substratum's physical form: the inner expansion.

What are the proceedings of this other modality of expansion? Now, what is at play is a new target to be deciphered: the mind. Mind also filled with images that are formed no longer in an external plane, but in seven inner screens or mental layers stratified under the title of *kos'as*—from the Sanskrit "layers": ***annamaya kos'a, kamamaya kos'a, manomaya kos'a, atimánasa kos'a, vijinánamaya kos'a, hiranyamaya kos'a, and átman***. In this sense, to meditate would be the constant effort of taking distance in order to better grasp the *átman*: the soul where the inner "Self," immortal and divine, dwells. But before reaching it, between "I" and "God," "me" and "soul", there is a long path to be taken: one must leave the air through which the wireless waves wander and unite the seven layers of the individual mind with the waves of the Cosmic Mind's oceanic depths. Let us prepare our lungs, let us grab our diving gear and take a deep breath: we will go from the impulsive longing to mediate to the disciplined structure of meditation. We will move on, exfoliating and penetrating layer by layer, plane by plane, but at this initial stage, we will focus on the first layer: *annamaya kos'a*.

What exactly is the most superficial of the mind's seven screens related to? "*Annamaya kos'a*" literally means "layer made of food": it is the body we see, touch, grab, and feel as something that belongs to us, defines us, and makes us who we are. It is also the first layer to be broken by Tantra Yoga: "the fundamental consciousness to the spiritual development is the realization that 'I am not this body'" (Ánandamitrá, 1999, p. 8). But then, what would the body be in this thing that it is not? Facial features, corporeal outlines, pores, skin, mouth, tongue, nose, ears, sex, and all secretions, saliva, and affections that emerge from such sensors through which we project ourselves, make an image and turn it into a profile, are summarized in a single and categorical phrase: "the human body is a biological machine" (Ánandamúrti, 2005, p. 93). By denying the identification with the palpable concreteness and with the visible image of the body, we are incited to search for its mechanisms: mechanisms that turn the body into a material vessel employed by the mind in order to operate in the physical world (Ánandamitrá, 1999). A model that is no longer narcissistic, reflecting the "I" through an external image on the plane, but a sensorial and empirical model that peels the skin and invites us to enter what is in the guts: in the circuits that intertwine nervous system and endocrine glands with the subtle anatomy of the "*cakras.*" It is up to the spiritual newcomer, then, to utilize the yogi tools to study and dissect how these cogwheels operate in the corporeal device. Such knowledge is the first step in the search for stopping the unceasing cycle of pleasure and pain, comfort and anguish, happiness and sadness that is experienced by the body in the inevitable collision between bodies with which we interact during our daily lives. Thus, by turning the plane of the *annamaya kos'a,* of the physical body, into an adjustable apparatus, Tantra Yoga poses a radical and seducing question: what would be the conditions of possibility for the affections of the biological substratum?

To answer such question, we must use the tantric scalpel to cut the skin and open the enigmatic connections between these three systems: endocrine glands, nerves, and *cakras.* Beginning with a few basic definitions: (a) endocrine glands are epithelial organs responsible for gushing into the blood stream hormones that circulate inside the body; (b) nerves, on the other hand, are axonic extensions responsible for conducing electric impulses coming from all corners of the body toward the central nervous system and, from the central nervous system, they transmit the commands generated there back to every part of the body; and finally, (c) *cakras*—from the Sanskrit "circle"—are energetic vortexes that work as substations of the mind distributed across seven points in the spine. All beings possessing an endocrine system have nerves and, consequently, *cakras.* Newts, frogs, snakes, rats, mosquitoes, bees, spiders, whales, octopuses, kangaroos, cats, cows, horses, and dogs. In human beings, the location of such glands inside the body coincides with the exact location of the confluence between three energy channels—the *nadiis id'a*, *sus'umna*, and *piungala*—that host the energetic vortexes distributed in the spine. A new species of knot: instead of the intersection between profile dots on social networks, what is at play here is the weaving of the glands/*cakras* joint throughout the neural networks' energetic points of *annamaya kos'a*. This constitutes the "*eureka moment*" of the yogi analytics: to map the production of affections emerging from the bond between the chemical cocktail of hormones and the vibrational operations of the seven *cakras* (Fig. 3.1).

Outside the traditional medical corpus, according to Tantra's Biopsychology (Ánandamúrti, 2005), any kind of affection is nothing more than a hormonal torrent transformed into electrical impulses to be controlled by the continuous regulation of *cakras*' points. Regulated in a way so that the images that are thought do not propagate in the nerves the mental vibrations of gluttony, greed, lust, wrath, envy, sloth, and pride—the seven deadly sins by Christianity that synthesize all the possible

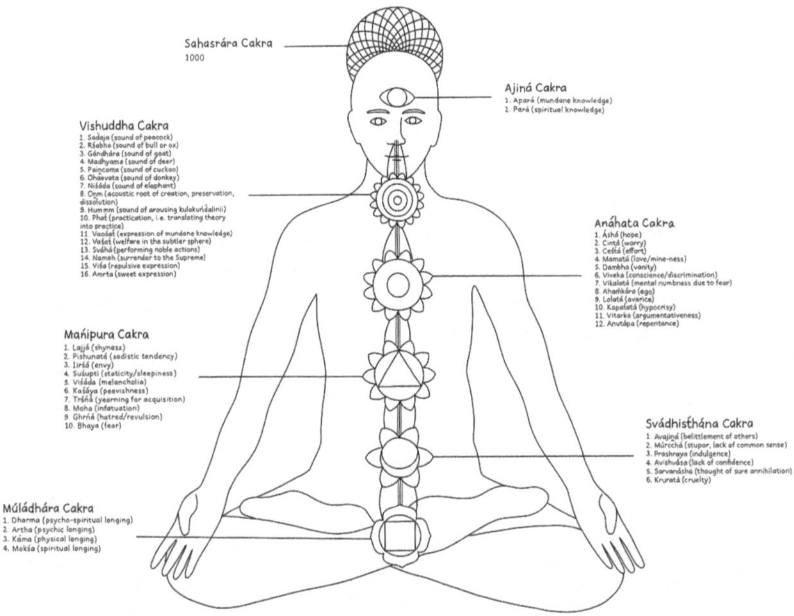

Fig. 3.1 The biopsychology of cakras and vrttis, from Tantra Yoga by the author

affections in the *vrttis* of the so-called lower *cakras.* "*Vrttis,*" from the Sanskrit "whirl", are like petals of a flower that are similar to small springs that, when vibrating, manifest a certain psychic propensity. Instead of six affections programmed by Facebook in the form of the emojis "Like," "Love," "Haha," "Wow," "Sad," and "Angry," now it is from the sum of the fifty petals distributed throughout the *cakras*' seven flowers that each and every affection of the human mind will emerge. However, there is another difference: the energetic points of *annamaya kos'a* are no longer located on the surface of screens that glow with images accessible to the mere touch of a fingertip. Burrowed inside the body, such dots capable of controlling the mind's "lesser" tendencies are only touched through two main methods of subtilization of the biological substratum waves: the complete twitching of the body through the internal pressure applied by *ásanas* and the modification of the cellular density provided by following a *sáttvika* diet.

We must then put the smartphone aside, forget for a while the astonishment provoked by the topology of Facebook's twitching dots and start creating malleability in another space: the space occupied by the solidity of the body. It is no coincidence that one of the basic rules of etiquette in any yoga class is that we must turn off or silence such devices. I confess that such demand was never pleasant to me: to stop being socially available to all profiles on the network at first caused in me something similar to an itch. Nevertheless, my spine would bend with guilt, my heart would harden with sadness, my belly would burn with anger, and my head would hurt from all the overthinking I was doing: I needed to fix that. The solution was, then, to disconnect from the outside and to connect with the inside: to touch the *cakras* points and minimally control the sea of hormones that were inciting my *vrttis* to express some unpleasant emotions. To that end, for at least one hour I would give up the frenzy of my fingertips in order to integrate bones, muscles, tissues, and tendons into a daily routine of bending, twisting, and contorting my body using *ásana* positions (Fig. 3.2).

"*Ásana*" literally means "position in which one feels comfortable" (Ánandamúrti, 2008, p. 23). Not all of them, though—and because of it there is a high risk of some ásanas to become flamboyant selfies shared as a form of social validation of the vigorous physical effort employed. But in general, yes, they are accessible positions that do not pose a threat to the strength of our muscles or to the laws of gravity: they consist only in slow movements accompanied by deep breathing and, alternated by periods of complete immobility, creating a deep relaxation of muscles and nerves (Ánandamitrá, 1999). Inhaling and exhaling. Pulling and pushing. In these cycles of actions and rhythmic fixity, the air slowly circulates between limbs, torso, and head and carries out the cellular oxygenation necessary to the distribution of vital energy (*pranah)* across the body. And by regularly synchronizing the inhaling and the exhaling to the twistings of the body, this process is not only contributing to the muscle elasticity and invigoration but, above all, it pressures the *cakras* points where the endocrine glands are located. This is the main effect to be achieved through the practice of *ásanas*: "to help calming and controlling the mind" (Ánandamitrá, 1999, p. 54) by means of compressing and extending these nodes responsible for the hormonal gushing that composes the sensitive tissue of the body. Thus, among the 50,000 possibilities, some positions became common in the daily effort to regulate my affections: *yoga mudrá* (yoga pose), *dirgha pran'áma* (long

Fig. 3.2 Ásana postures, by Tantra Yoga

salutation), *gomukásana* (cow head), *janushirshásana* (head on the knee), *naokásana* (boat pose), *bhujangásana* (serpent pose), *matsyendrásana* (twisting pose), *padahastásana* (hands on the feet pose), *sarvangásana* (candle pose), *matsyamudra* (fish pose), *halásana* (plow pose), and *shasangásana* (rabbit pose).

By mimicking the images of animals, objects, salutations, and angular movements, the disciplined repetition of *ásanas* provides access to an interaction with the points hidden in the electronic circuits of the *annamaya kos'a*'s physical plane. Trine circuits, comprised of the combination of nervous endings capable of expressing in emotions the hormonal information secreted by endocrine glands that, in turn, are subordinated to the controlling points of the seven *cakras*. Electrons, liquids, and vibrations: three entities that organize themselves within the biological substratum to process the energetic circulation of waves across the body, or in yogi language, the *pranah*. And it would be to no avail the effort of twisting and turning *annamaya kos'a* into *ásanas* to touch its cakra points while, at the same time, charging the batteries of these circuits with food coming from dense energies. Hence, in the package of the desire for meditation is embedded the possibility to alter the sources from which the pranic sap is extracted. Can you imagine eating a hearty "*feijoada*" (Brazilian black bean stew), performing *ásanas* and then, right after that, sitting down to meditate? I would not say that such combination is impossible, but it probably would result in a stodgy situation. This is the case because, every time we eat, a good amount of blood is concentrated in the abdominal region so that the ingested food can be slowly digested and converted into energetic fuel for our corporeal device. Logically, the heavier the meal, the longer the intestinal absorption will take and, consequently, the higher the toxicity index generated by the food in the body. To avoid such inconveniences, a second method of subtilization of the biological substratum is made available by Tantra Yoga. This method is the practice of the so-called *sáttvika diet*.

How I wish we were like smartphones that can receive pure energy through a simple cable—energy that comes already chewed and electrified—to charge their batteries. But no. We have mouth, guts, anus, and a complete digestive system full of teeth, liquids, and enzymes that are in charge of grinding everything we put inside our bodies until we can finally obtain the energetic juice that fuels the body. It is through this prism that the *sáttvika* diet emerges as a complementary technique to the practice of *ásanas*: it works as an energy extraction tool arising from the ingestion of a certain group of foods that makes the body "lighter," that is, less overwhelmed by the process of digesting denser frequencies of matter. If there is the option of charging the phone's battery with energy coming from cleaner and renewable sources, and no longer from burning coal at thermoelectric plants, why not do it? A similar logic can be applied to food: starting from the presumption that food's vibrational quality directly affects the physical and mental vibrations of those who eat it, the *sáttvika* diet avoids food coming from the death of beings with a complex nervous system, and from hallucinogen or hyper-stimulant sources. Hence, the *sáttvika's* nutrition group is quite specific and, in general, cannot be found at shopping centers' food courts: it is restricted to a vegetarian diet, filled with grains, nuts, legumes, fruits, milk and dairy products, and vegetables with the exception of garlic, onion, and mushrooms. No eggs as well. And what is the reason for that? If the body becomes more subtle, and less charged with the energy originated from animal suffering and psychoactive substances, the same will happen to your mind, which slowly will become less agitated and disturbed.

It is no coincidence that the *annamaya kos'a* is literally the "layer made of food." By taking seriously the old saying that goes "you are what you eat," the food substratum is classified not based on aspects like the taste or nutritional properties: what is taken into account is the vibrational quality of the material in question.[3] And if the Cosmic Mind vibrates on a subtle wave frequency, those who want to move away from a series of socially ingrained habits to better grasp it via the mediation of the mind need to, in order to touch such Mind, go through a process of subtilization. Drastic distance. To subscribe to the sáttvika diet was a little unpleasant to many people in my circle of social interaction: being the daughter of two

[3] The categorization of *sáttvika* and two other diets, the *rajásika* and the *tamásika*, are closely related to the creating forces of the universe: the cosmic cycle or "*brahmacakra*." As the name says, *brahmacakra* literally means the "circle of Brahma," an infinite entity, eternal, and immutable that contains (and is present in) everything. And according to the cosmology of Tantra Yoga, Brahma is expressed by the conjugation of two principles: the cognitive principle of *Purusa*, a consciousness that observes the universe; and the operative principle of *Prakrti*, which works as a linking energy under the command of *Purusa*. Whereas *Purusa* placidly observes, *Prakrti* is the energy that does all the heavy lifting and acts on the densification of the Supreme Mind into palpable material. In this process, *Prakrti* utilizes three distinct attributes of force—the *gun'as*: the subtle force or *sattvagun'a*; the mutative force or *rajogun'a*; and the static force or *tamogun'a*. It is from the qualities of each *gun'a* that food is distributed into three general categories: *sáttvika* food, linked to the subtle force; *rájasika* food, related to the mutative force; and *tamásika* food, associated with the static force.

carnivorous "*gaúchos*[4]" who own a chain of Italian restaurants filled with delicious dishes marinated in high doses of garlic and onion, by proclaiming to be "vegetarian" and by resisting such spices, all of a sudden I became the "alien" of the family. Questions such "are you sure this is healthy?" became a frequent source of worry to my mother—and also a target for constant questionings from some friends who were used to the Camila who ate steak with onions and would drink a few beers at the bar every once in a while. French fries became the snack of choice and soda water with lime and ice, the favorite drink. Nevertheless, if in the mediation practices, the meeting between body and device produces the introjection of energy through the touch of fingertips that feed us with images, what is now at play is another degree of material rarefaction. By turning the own body into a device, a means to connect with the divine, the vibrational information that composes the food and is ingested by the mouth, penetrates deeply. As deep as the *cakras'* energetic centers that regulate the production of images in the mind. And who knows? Maybe someday we will be able to perform photosynthesis and feed ourselves solely via the consumption of light. Until this day comes, what is left to meditation newcomers is to proceed with the proper subtilization of themselves in order to experience a new stage: to sit in lotus position and dive even deeper into the affections of other planes or layers of the mind.

3.2 Mind and Points

Imagine a circle (Fig. 3.3). Imagine also that in the middle of this circle, there is a point. Attracted by the force of this minuscule yet so magnetic dot, the circle spins: an individual "I" dances around a Center, a Nucleus, a big "Self", called *Parama Purusa, Átman* or simply, God. How to grasp that divine point of the "Self" around which the "I" vulgarly spins? How can we merge the unitary "I" with the limitless "Self"? When facing the anguish of the distance that separates circle and nucleus, there is only one option left: to stop spinning around and move toward the inside, or what P. R. Sarkar called "to cut through the radius" (Ánandamúrti, 1979). Geometrically, the radius is the distance from the center to any point on a circumference. Meditatively, this implies the creation of a straight line, a passage, a channel that allows a direct path for the connection between circle and center, "God" and "I", "soul" and "me". That is the case

[4] A "*gaúcho(a)*" is everybody who was born in the Brazilian state of Rio Grande do Sul, famous fot its barbecue.

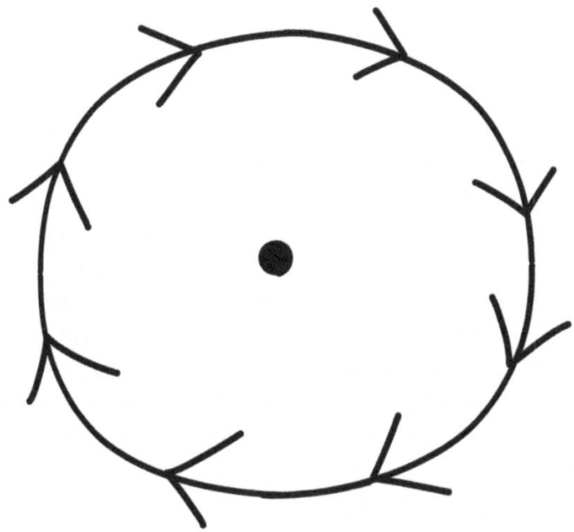

Fig. 3.3 The circle of the "I" and divine dot of "Self", by the author

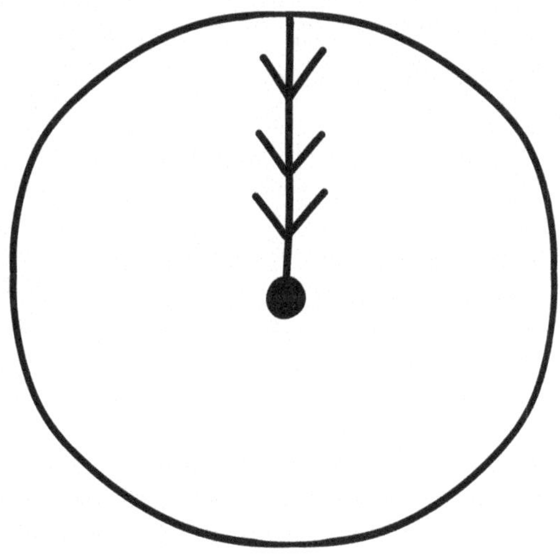

Fig. 3.4 "To cut through the radius", by the author

because, after having prepared the physical body's vibrations with *ásanas* and the *sáttvika* energy, a new stage of subtilization takes place: the transformation of all mind's images into one tiny dot small enough to go through this narrow and long shortcut created by the method of "meditation" (Fig. 3.4).

If by mediating, we create devices of technical image that today exteriorize the "I" through planes in the form of energized profile points, to meditate involves another kind of technical procedure: the shutting down of every plane of *kos'a* via the progressive introversion of the mental device's working mechanisms. To that end, it is time to rarefy the next *kos'a* better known as "*kamamaya kos'a*." "*Kama*" comes from the Sanskrit "desire." Consequently, *kamamaya kos'a* literally means "the layer of desire." This is the first plane of the conscious mind, responsible for controlling the previous layer (*annamaya kos'a*) and mobilized by an incessant search: from pleasure to pleasure, from desire to desire, *kamamaya kos'a* wanders the external world searching for a new object of satisfaction. A misgoverned search, hungry for the continuing excitation of the ten *indriyas* or sensory-motor organs: five organs of knowledge or *jinánendriyas*—ears, eyes, skin, nose, and tongue—and five organs of action or *karmaendriyas*—hands, feet, vocal cords, excretory and genital organs. It is by the combination of these ten sensors attached to the plane of *kamamaya kos'a* that one is capable of performing three crucial functions: (a) to feel, (b) to respond with aversion or attraction to the received stimuli, and (c) to act when facing the attraction or aversion input. Functions that are closely related to the four basic instincts of hunger, sleep, fear, and sex in service of our animal auto-preservation and reproduction (Ánandamitrá, 1999): have you ever thought about the dangers of not reacting when facing the feeling of pain or giving up the search for pleasure? However, in the human animal such instincts are manifested in a sophisticated way. The necessity to eat goes beyond being satiated, in the same way that sleeping, feeling fear, and having sex are more than a mere vital maintenance: consciously, we desire.

Hence, it is no coincidence that the Upanishads[5] affirm that the "mind is a wagon pulled by ten wild horses." In a world of hyperstimuli,[6] especially advertisement stimuli that incite us toward delight and satisfaction through the body's continued electrocution on screens, we must say "no more." Enough of that indomitable overload of impulses and vibrations

[5] The Upanishads integrate the Vedas and can be found among the texts in Old Sanskrit that contain some of the fundamental philosophical concepts of Hinduism.

[6] According to Singer (2010), today we are experiencing a different level of stimuli: the hyperstimuli of the so-called neurological modernity (Singer, 2010), which entails other transformations in the structure of the experience. In it, the "intensification of the neural stimulation and the physical liability" (p. 117) are the touchstones to the composition of other ways of feeling and dealing with the contemporary sensorial regime.

that provoke the *kamamaya kos'a* to search for more and more pleasure. "Enough" that ends up with an abrupt intervention: to end the spinning of the circle of the "I" via the complete paralysis of all *indriyas*. Beginning with the motor organs: instead of gliding the fingers on planes, the extremities of both hands would be fixedly intertwined on the lap; instead of moving the legs and laying down, running, walking, or stopping in a particular way, they would be crossed in *Padma'sana* (Lotus position) and, in one fell swoop immobilizing feet, genitals, and excretory organs; instead of speaking, the vocal cords would be silent in order to hear only our inner noises. Thus, it is done: hands, feet, vocal cords, excretory and genital organs are now locked without any possibility of moving whatsoever. The next step, then, is to remove the sensory organs: by closing the eyes, vision is shut off from the smartphone screen; the tongue is folded on the palatal dome in order to control the gustatory activity (Ánandamúrti, 2008); the fingers of intertwined hands and the feet crossed in lotus help in the progressive shutting down of the touch sense; as the ear cannot stop hearing, the solution is to protect it by going to a place free of noises, clatter, and mainly, free from the rowdy interruptions of cellphones; finally, the sense of smell becomes isolated in a fresh environment, without the presence of strong odors or incense.

When all *indiyas* are encapsulated, one can finally utter that simple phrase, better known as "sitting down to meditate." It is as if we would grab the smartphone and remove its antennas, its earphones, and its seducing screen: what is left of this act of sitting down? What happens after the hardening of the spine and the deactivation of nearly all sensors responsible for the "I's" connection to the external world? Individually, it is up to each person to test it for themselves: to accommodate the legs as best as possible and to get in touch with the singularity of your own experience. But, in general, there is indeed a common and shared living experience. By deactivating the cogwheels that activate the production of desire, a thousand images invade the mind: thoughts, memories, planes, visions ... That family problem, which time and time again comes back to create yet another conflict, the bills that are still to come, the belated wage, the latest post on social media, the movie one saw last week, the news about the moral bankruptcy of politics, the work that depends on the goodwill of so many other people to get finished, the meeting with friends at a bar, the delicious Thai food, the philosophical ideas read in a book, that touching poetry, and so on and so forth. And another detail: in general, these images do not stop. On the contrary, they are so intense that they torment us.

Should not it be the exact opposite? What about those portraits of wise yogi masters and monks meditating with the most placid expression of peace on their faces?

It is hard to find peace when one enters once and for all the subconscious plane of *manomaya kos'a* or, literally, "layer of deep thought." "Once and for all" because normally its activity is eclipsed by the frenetic activity of the *kamamaya kos'a*. Nevertheless, temporarily stunted by the mechanisms of attraction and aversion toward the outside as a result of being in a meditation position, the desires that were subconsciously repressed all of a sudden emerge. Do not fall into despair if you took the risk of intertwining your hands, sitting down in lotus position, closing your eyes, isolating your nose and ears, and all kinds of thoughts are popping up on your mental screen. The images that will appear are neither in vain nor exist only to perturb you. Their only goal is to perform three functions: (a) to experience pleasure and pain via thoughts; (b) to think deeply through dreams; and (c) to reflect by means of the use of memory and intellect. This implies that, without the *manomaya kos'a*, it would be impossible to solve our daily life problems, to think philosophically, to process experience through dreams, to remember and recall events, and to manage information on an intellectual level. However, in the same way as it happens to the body, this torrent of vibrations that invades the mind is also an obstacle to the desire for limitlessness: the absurd quantity of images that pops up in the plane of this *kos'a* is obviously not small enough to go through the microscopic passage that connects the circular "I" to the divine point of the "Self." Thus, another technic of tantric intervention becomes necessary: if to sit down to meditate allowed for the spinning of the "I's" circle to stop, one must now disintegrate the linearity of the circle into points and reduce them to a single dot. But what exactly does this process mean? (Fig. 3.5).

The Euclidean geometric shape of the circle has quite peculiar characteristics: it does not have a vortex, a base, or any tips; it has an inner area, a center point, and a radius whose distance is always the same. The circle is so round that it is self-sufficient: the circle is closed in itself. To shatter such solidity of the circular form is not easy, but it is inescapable for the act of meditating. That is the reason why the circle of the "I" has to crumble and lose the boundaries of its outlines. And such process is performed through the so-called *shuddhis*, from the Sanskrit "purification." In a total of three, the *shuddhis* are visualization techniques that put into practice a process of suspension. It means the mind's extraction from the

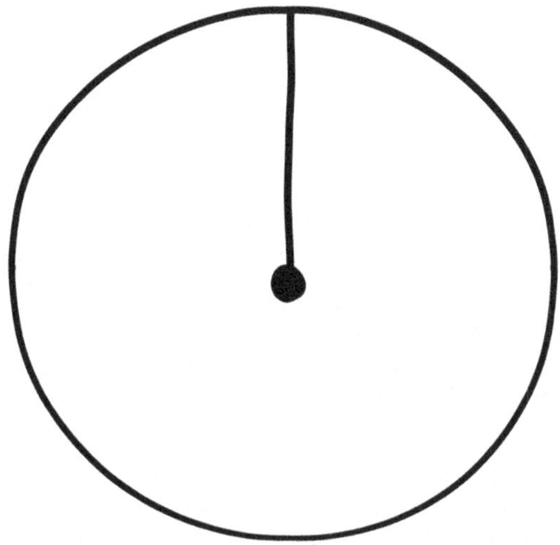

Fig. 3.5 Paralyzing the "I" circle spinning, by the author

spatial-temporal limits so that it can assume the minuscule existence of a dot. It begins by the removal of the external environment with the first *shuddhi*, the *bhuta shuddhi*. "*Bhuta*" means "created object," that is, everything that has an existence to the senses and bumps into the physical body or *annamaya kos'a*. In this regard, *bhuta shuddhi* aims to achieve the complete shutting down of the external space in which the body is inserted: it is the "process of mental abstraction in relation to the external world" (Ánandamúrti, 2008, p. 146). In other words, to move away from the concrete in order to better grasp it via visualization techniques or internal images that are seen even with eyes closed in order to "remove all external vibrations from our mind" (Ánandamitrá, 1999, p. 66). No longer the Facebook polis with its topological space filled with pictures to be seen and touched or the tridimensional space in which we wander across the natural and urban landscapes. Simply, there is no longer any experience of the physical space: it doesn't matter if the roundness of the "I" has sit down to meditate in the silence of the Himalayans or in the middle of the urban chaos, by applying this technique, none of the sensorial stimuli arising from the external environment should be capable of producing that which is sensitive or sensorial.

The circle of the "I" is thus isolated: outside the living room, outside the house, outside the city, outside the state, outside the country, outside the continent, outside the planet. There, fatally alone, it hovers above the openness of the cosmic space. Only the blackness of the cosmic immensity witnesses that which became the only existing entity among everything there is. It is at this moment that begins the second *shuddhi*, the "*ásana shuddhi*," responsible for undoing "the conditioning that caused the mind to identify with the body" (Ánandamúrti, 2008, p. 146). To Westerners, this is one of the hardest blows delivered by the yogi toolbox. A precise and swift hammer blow. From it, the only thing that remains is the dust from the fluctuating points of what was experienced as everything we are: a closed circle possessing an individual body. From the Latin "*individuus*," that is, undivided. Concrete, recognizable, and unitary. Consequently: egotistical and self-sufficient. To shatter such identification with the self-image, with selfies, and with pictures posted on the profile, hurts. Just like an egg whose shell breaks and what remains is only a shapeless and slippery gelatin, everything that we worship, wear, undress, comb, beautify, and that every day the mirror shows as a reflex of the "I," suddenly comes crashing down. It does not matter how beautiful or ugly the eyes consider the epidermal cover to be, below it there will be a disfigured goo that, pertaining to the "I's" disintegration circle, relates to a goop of dots that wanders the subconscious mud of the *manomaya kos'a* (Fig. 3.6). A mud that is dense, filled with projections, fears, doubts, and searches: mental

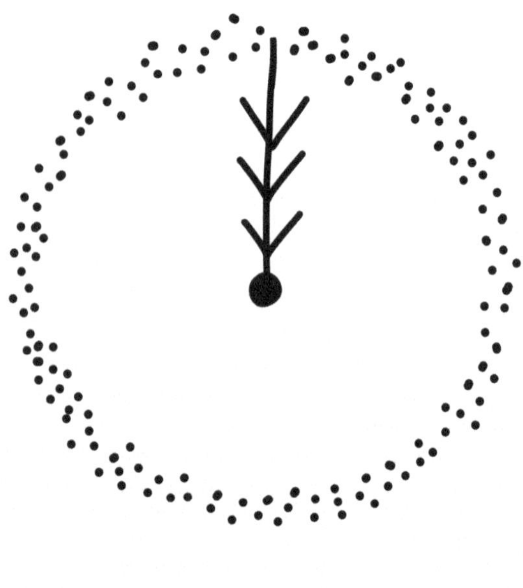

Fig. 3.6 The fragmentation of the circle of the "I", by the author

images. Hence, after the grinding of the "I's" circular existence, which is identified with the corporeal image, we must keep macerating the mind so that the dots of its images are reduced to the tiniest quantity: a single and singular point in the universe "as focused as the tip of a needle" (Ánandamúrti, 2001, p. 1).

How can we do this? Would it be possible for the mind to become as small as a miserable dot? Just like a laser that increases its potency by concentrating itself on a single frequency, the mind must work in the same way: a "single-pointed mind" or "*agryábuddhi*." However, to reach such state, in which "the mental mirror becomes pure" (Ánandamúrti, 2001, p. 1), one must not only remove the body from the physical space but also erase the corporeal existence's concreteness through the execution of the *ásana shuddhi*: again by means of using visualizations, now the sum of all dots that inhabit the mind go through an intense process of funneling (Fig. 3.7). A disciplinary funnel: if in the act of mediating, to touch the points of image represents an automatic gesture that requires the bare minimum of effort from the fingertips, in the act of meditating what happens is the complete opposite. Only through a series of techniques of continuous effort that such endeavor can be achieved. It is no coincidence that the Sanskrit word for meditation, "*sadhana*," literally means "effort." It is in this hardworking exercise of the mind that the attention is gradually

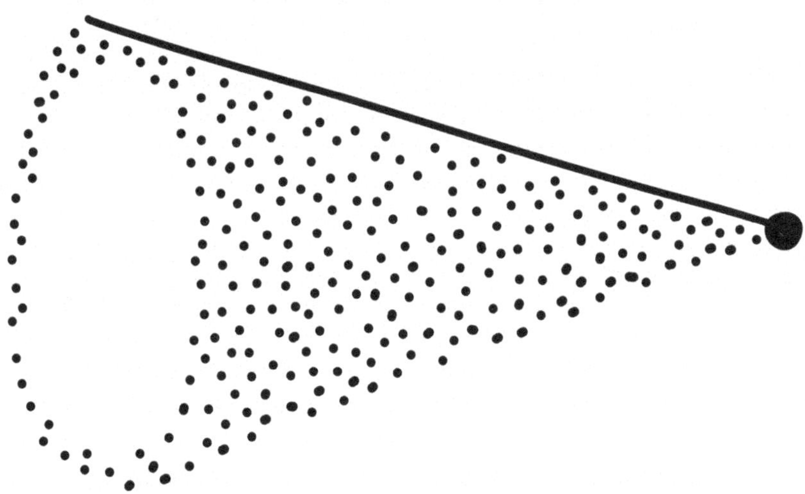

Fig. 3.7 Funneling of mental dots into a single dot, by the author

shut down "in every part of the body and is concentrated on a point known as *Ista cakra*" (Ánandamúrti, 2008, p. 146). Here, the space occupied by the biological substratum no longer exists: the "I" simply becomes a mind manifested in the form of a single dot. And properly pointed out, it will fixedly be located in the *Ista cakra*[7] in order to find there a definitive home for the dot, determined by one among three possible *cakras*: *anahata*, *vishuddha*, and *ajina*. There is no longer the possibility of exchanging dots through the shock of the fingers on smartphone screens or of sending them to the most remote places via Wi-Fi waves. The dot of the "I" is rigidly cemented so that the mind stops wandering and starts concentrating only there, standing still in that small blurred image.

At this stage of meditation, head, torso, limbs, organs, muscles, bones, tendons, nerves, secretions, and body senses are suspended: the outlines that form the human shape are gone. With our eyes closed, everything that we see in the darkness of the cosmic space is epitomized in this lonely dot that inhabits a *cakra* point. Lonely, but not too lonely. The "I's" point wants to grasp the Divine "Self's" Dot, the Supreme Mind beyond the restrictions of space and time. The "I" longs to become unlimited. To that end, the third and last *shuddhi* has to take place: the "*citta shuddhi*." "*Citta*" designates the "mental substance" or "ectoplasm." In it we have a new target to be attacked by the tantric scalpel: since everything that is left of consciousness is no longer in the external space or in the body, but only in the mind, or rather in the tiny dot that it has, we have to dissect the constitution of this dot in order to make it even smaller and sharper. Sharper, that is, more subtle and, therefore, more capable of getting closer to the divine and inner "Self." Hence, we must suspend the mind's densest portion, which is responsible for the production of images: *citta*. It is not by chance that *citta* is considered as the "mental screen" (Ánandamúrti, 2008), in other words, the part of the mind that assumes the shape of the perceived sensorial vibration or of the planned action. Yes, it might seem a tad weird to our "Westernness," but according to tantric yogis, there is a portion of the mind that is literally transmuted into external objects based on everything we see, hear, taste, smell, or touch and, in a nutshell, on everything that we think or receive as external stimuli. Imagine how many images are processed in the scrolling of the newsfeed? It is this hyperactive

[7] Each person has their own personalized *Ista cakra*, orally designated and in a presential manner by a proficient instructor of meditation. Such choice will depend on the frequency of each personal vibrational rhythm.

fraction that we must now stop in order for the "I" to be able to think one single ideation: the image of the Cosmic Consciousness.

How to train *citta* so that it stops transmuting into the form of all planes, doubts, certainties, desires, fears, ideas, instincts, and external stimuli that assail it? The contradictory fact is that to achieve such goal, one must not bother trying to stop thinking. It is simply "impossible to think about nothing. The mind always has to have an object" (Ánandamitrá, 1999, p. 69). You can try it for yourself. When it seems that you are thinking about "nothing," "nothing" becomes your new object of thinking. And at the moment somebody else suggests: "do not" think about the blue elephant, there it will be, immense and unflinching, exhibiting its blueness. Thus, in order to end the activity of *citta*, *citta shuddhi* proposes a third visualization technique: to imagine that the dot located on the *Ista cakra* point is immersed "in the ocean of the Cosmic Consciousness" (Ánandamúrti, 2008, p. 146). By projecting the image of these waves submerging and encompassing the "I" in the cosmic sea, *citta* is then dissolved into another and more subtle part of the mind called "*aham*" or ego. There, in *aham*, resides the feeling "I do": in it we feel both the authorship of the act of making and the feeling of possessing the actions performed by the "I." "I do" which, when removed from the images of external objects and immersed in the comings and goings of the divine waters, gradually reduces even more the size of the "I's" dot until it becomes sufficiently small to, at last, go through the narrow tunnel of the radius that leads directly to the Nucleus of *Parama Purusa*. What is expecting us in this new stage of intersection? Will it be possible for the "I's" dot to traverse the channel conducing to the *átman*? (Fig. 3.8)

3.3 Mind and Shocks

From clash comes cohesion. (P. R. Sarkar)

Finally, after three *shuddhis* we were able to let go of three distinct dimensions of experience: our surroundings, our body, and the torrent of subconscious thoughts. From the circle that could only curl and was disintegrated into a huddle of misgoverned dots, a single and lonely dot of the "I" remained, ready to go through its last and final path toward the divine dot of the "Self": the three final layers of the mind—*atimanasa kos'a, vijinánamaya kos'a*, and *hiranyamaya kos'a*—that take us to the *átman* or "soul" (Fig. 3.9). For those who thought that this would be a

Fig. 3.8 The duct between "átman" and "I", by the author

smooth journey, they are wrong: there is no way to cross the finish line without experiencing an innumerable series of shocks or clashes stored in the first layer of the superconscious mind, "*atimanasa kos'a*" or, literally, supramental layer. But what kind of shock are we talking about? And how is it possible for "shocks" to be stored in one of the planes of the human mind? Well, some of the thoughts, images, and dreams of *manomaya kos'a*, those paranoias and fears that seem irrational were already a clue, signs that at some point such clashes will in fact be manifested. Very well, here they are. Shocks that present us situations that are not always pleasant or capable of being liked and shared on social media profiles: the death of a close friend or relative, a painful romantic breakup, being fired, an accident, false accusations, financial problems, a sudden and unexpected robbery, existential doubts, a deep depression… Anyway, all kinds of misfortunes—or fortunes as well—that may happen are nothing more than tantric shocks expressed in the form of "*samskaras.*"

However, before looking into this process in further detail, what exactly are "*samskaras*"? In general, *samskaras* are potential reactions. In them, the maxim of Physics that affirms that to every action there is an equal and opposite reaction is taken to its last consequences: in Tantra, each and every action has not only "external reactions, but also internal ones, that

Fig. 3.9 Crossing the last three layers of the mind, by the author

is, the author of a conscious action also suffers an internal reaction" (Ánandamúrti, 2008, p. 131). Such premise implies that "it doesn't matter if there is a physical action or a behavioral expression, if there is a thought, an action (*karma*) will be performed" (idem). Expressions like "that son of a bitch, one day he will pay for this," "what the hell is this person doing?", "idiot!", "what an ugly woman!", "when will this moron wake up to life?" Yes, all of these narrow-minded and petty thoughts are actions. And since the source of all *karma* is the *aham* or ego, the psychic effects of the performed actions will also be experienced in it. Actions that might not move arms and legs, torso and head but, by moving wave frequencies that create vibrations in the distinct planes of the mind, they immediately produce, as it happens in Physics, an equal and opposite reaction. Equal and opposite, but this does not mean that such symmetry will take place through the repetition of the same action or experience that has engendered it. It is that old saying that goes "live by the sword, die by the sword[8]": meaning that the reaction will take place on the same level and intensity as the action that produced it, even though it might take place in a different way or in another incarnation.

[8] In Portuguese, the saying would go as: "*Aqui se faz, aqui se paga*" (Here you do, here you pay).

In Tantra, there are three kinds of *samskaras* or possible reactions: (a) innate; (b) imposed; and (c) acquired. Innate because they are related to the "mental tendencies inherited from past lives that shape the physical bodies and the minds of the individuals in their current life" (Ánandamúrti, 2008, p. 133) and that are translated into phobias and inexplicable preferences, innate diseases, or extraordinary abilities, for instance. Imposed due to the fact that they are related to social factors such as "the culture and the environment" (idem) that shape the personality and subjectify moral attitudes such as racism, sexism, nationalism, and many other daily "isms." Finally, acquired *samskaras* are those produced during the period of one's life arising from "actions performed by people having a feeling of ego" (Ánandamúrti, 2008, p. 134). From the combination of these three kinds of *samskaras*, emerges an "I" full of *vrttis* or psychic propensities that will act and, with each performed action, will leave an impression, a mark in the mind. It is hard to escape unscathed and to have a smooth mind, without any marks or scars as a consequence of our actions. And little by little, these marks start to pressure the mind just like a ball full of air is disfigured by the fingers when an external pressure is applied. The screen of the mind is then deformed by the impressions of *samskaras* and wants more and more to go back to being a screen as smooth as the smartphone's touchscreen. It is at this moment that the *samskaras* are deflagrated: the potential reactions that are prevented "for long periods of time until the circumstances of the environment are propitious for a reaction" (Ánandamúrti, 2008, p. 132) are finally actualized into events that simply aim to repair such mental distortions.

To navigate *atimanasa kos'a* can be a treacherous endeavor. At the same time as it is the plane that opens the doors to the realms of intuition, imagination, creative insight, premonitory dreams, clairvoyance, and telepathic communication, this plane also faces us with a challenge. The dangers of *aham*, the ego that, by claiming for itself the authorship of the action, falls into the risk of turning such abilities into "superpowers", multiplying even more *samskaras*. *Aham*: a remnant of *citta shuddhi* that now constitutes the new obstacle to be shattered by the continuous mental effort of meditation. Which new techniques will be necessary so that the "I's" dot does not become stagnated in traumas and spells and, thus, is not able to complete the race toward the *átman*? To undensify the mind even more, and fusing *aham* into *mahat*, that is, intellect into intuition, Tantra possesses yet another tool: the use of a "*mantra*" or, literally, "that which liberates the mind". At this moment, the point of the "I" fixated into the *Ista cakra* and immersed in the waves of the Cosmic Wave also becomes a vibrating dot. It vibrates thanks to the communion between three qualities: a mantra that is (a) pulsative, (b) incantative, and (c) ideative. Pulsative because it is composed by the pulse of two syllables: one mentally uttered during the inhaling, other during the exhaling. "Mentally uttered" since it is not recommended for the vocal cords to pronounce such sounds. One will simply utter in one's head—without posting any sort of chatter whatsoever on any corner of social media—the same melody: a duo of sounds extracted from the Sanskrit alphabet. This is the case because, besides communicating, the Sanskrit is an alphabet designed to perform an intriguing function: to harmonize the vibrations of the fifty *vrttis* located in the energetic centers of the *cakras*. How is that possible? Mother of all Indo-European languages, the Sanskrit alphabet was created from the profound meditation of yogis who, in the closure of the Himalayans, have clearly heard the sound produced by the *vrttis* of each *cakra*. Thus, by reciting a mantra, one is not only reciting a huddle of letters but also sounds that vibrate and pulse in order to enchant the "I", elevating its "personal vibration" or "entitative rhythm" (Ánandamitrá, 1999, p. 73).

From that comes the second attribute of a mantra, the incantative. Remember those images of Naja snakes that would stand erect thanks to the hypnotizing sound of a flute? There they are, enchanted, in a trance, enthralled by the communion between its vibrational pattern with the notes blown by a flutist who now commands the most venomous of all snakes. The charming function of the mantra acts in a similar way, however, the musician is the very own "I" and the serpent, in this case, is the *kúlakundálini* coiled around the first *cakra*. In order for it to go up, harmonizing the lower *cakras* and expanding the mind toward the unlimited, we must synchronize our entitative rhythm to the *mantra's* sound. A personal and singular mantra: *Ista mantra*.[9] In the same way that each and every animated or inanimate entity has their own entitative rhythm, each entity has a singular *mantra* that hypnotizes its own serpent and makes it to go up toward the divine dot of the Cosmic Mind. It is this entity, unflappable, immobile and, in absolute silence, that is the emanation

[9] The universal *mantra*, utilized before the initiation and personalization of the *Ista mantra* is the sonorous significant "*Baba*." Usually one of the first, if not the first, word to be uttered by babies, its meaning is "Love" and it works in the uttering of a "Ba" during the inhaling and another "ba" during the exhaling.

source of all the other wave frequencies that vibrate across the universe. An entity that is also the final destination of this lonely journey. And to reach it, we must persist even more strongly in the intense effort of concentrating on the mantra. Just like in a "*one note samba*,"[10] one has to exhaustively repeat a particular sound until the mantra and the entitative rhythm collide into the cosmic rhythm. A rhythm disconnected from the wireless waves generated by smartphone antennas that utilize radiofrequency to tune into as many Facebook profiles as possible. The connection here utilizes mental antennas, ready to be adjusted by the waves that access other states of consciousness beyond the individuated "I."

It is during this stage that the *mantra's* third and last property comes into play: its ideation. Besides synchronically pulsating both the sound and the inhaled air, and controlling the fifty affections of the *vrttis,* the *mantra's* sonorous signifier reverberates within itself a meaning to be remembered. An affirmation, an idealizing object intimately related to another yogi maxim: "you are what you think." And after all, what does the *mantra* suggest us to think we are? If we are not the body that delimitates us, or the senses that make us desire, or the logical thought and intellect of the subconscious mind, what does this sound tell us to be? Regardless of what is the semantic formation of the huge variety of possible *Istas mantras*, their phonetic always utters a single meaning: "I am Infinite Consciousness," "I am one with It " It is here, in this sublime meaning, that is hidden what the dotified "I," immersed and vibrating, so badly wants: the desire to expand itself beyond the limits of the physical body. The desire to be infinite, no longer restrained by the influence of space and time and by the deformation of the matter submitted to the force of gravity at all times. Desire to fly without the need of wings, to run at the speed of light, to simultaneously speak in every language, to exist beyond that which imposes limits to the biological substratum: to grasp the divine point of *Parama Purusa* and dwell in it without the need of images, without having the feeling of authorship, without intuition, and with no longer a body. Desire to simply be an undulating dot, reciting its divinity in the emptiness of cosmic space. In a nutshell: the desire to be one with everything.

[10] "One Note Samba" (*Samba de uma Nota Só*) is a Brazilian song by Tom Jobim and lyrics by Newton Mendonça, a classic of Bossa Nova also recorded by artists like Stan Getz and Charlie Byrd, Ono, and Stacey Kent.

"I am Infinite Consciousness." "I am Infinite Consciousness". "I am Infinite Consciousness". "I am Infinite Consciousness". "I am Infinite Consciousness". Ten minutes. Thirty. Two hours. Sixteen. The amount of time one intends to remain immobilized in lotus position reciting the sound of the mantra to the punctual "I" varies according to the limits and necessities of each person. "Necessities" because, far from being a relaxing process, throughout the practice of meditation one also experiences the shocks arising from the gradual exhausting of *samskaras* stored in *atimanasa kos'a*. This is the case because, if generally only in life and death situations or in states of deep coma the *samskaras* can become "ripe" (Ánandamúrti, 2008), during the tantric meditation they are literally "burned." Literally, it is not uncommon for the "I" to experience the feeling of being in flames during a meditation session. Without ornaments, pity, or compassion, the pain involved in such process can be cruelly atrocious: to get rid of *aham* is like entering a purgatory. Purgatory whose bonfire aims to perform an ecdysis, the shedding and changing of the skin of the *kúlakundálini*'s serpent wrapped in the scales of its own ego. In a raw and brutal way, affections arising from the most uncomfortable *vrttis* assail the mind while one is placidly sitting in lotus position. And even though the mind is suffering an intense pain, we must persist, keep applying the *mantra's* pulsative, incantative, and ideative effect due to a simple reason: the more *samskaras* we burn, the more the "I" advances in the narrow duct that brings it closer to the *átman*.

"I am Infinite Consciousness." "I am Infinite Consciousness." "I am Infinite Consciousness." "I am Infinite Consciousness." "I am Infinite Consciousness." Besides the heat, there is also another effect deriving from such insistence: the acceleration. Without further ado, life's events seem to spin at stratospheric speeds. What would appear to be stable, all of a sudden starts to corrode. What would seem to be concrete, is destroyed. Invariably, those who put themselves at risk in the bonfire of tantric meditation "experience their lives more intensely because they are ripening *samskaras* and consequently experiencing reactions at a high speed" (Ánandamúrti, 2008, p. 136). Speed of light that one wants to achieve so badly? It turns out that, without much logic and in short periods of time, the most unexpected personal turnarounds can take place. Turns that are no longer able to scroll the newsfeed wheel. The conversion of the mind into a means to grasp the soul will, instead of scrolling profiles incarnated on touchscreens, spin the cycle of birth and death of the *samsara* wheel. In this cyclic game of incarnation and reincarnation, while there are still remnants of deformations in the mind, the "I" will come to the world again and again until its potential reactions are completely exhausted. A wheel now engendered by the addition of past lives to the current one in the form of a singular "I" that is isolated from all the other dots of "I" in order to be merged with the divine point.

Thus, dizzy, bewildered, and clearly shaken by the intense spinning of *samskaras*, the mind finally is expanded and traverses *atimanasa kos'a*. It expands beyond the body's physical form and the individuated self, the ego and, through the correct use of the *Ista cakra* and the *Ista mantra*, it establishes the "I" in "*dhyana*", or literally, "the mind's flux." To flow. This is the last stage and the true meditation, and in it one crosses the layer of the mind that comes next: *vijinánamaya kos'a* or "layer of special knowledge." "Special" because, after so many highs and lows, of falling down and getting up so many times, the "I" finally learns how to not be affected by the shocks that assail it and starts to face those falls based on two affections: *viveka* and *vaeragya*, discernment and detachment. With *aham* properly dissolved into *mahat*, that is, with the feeling of ego transmuted into intuition, what is left is the crucial last stage: to immerse *mahat* in the *átman*. It is at this point that, freed from the illusion of being a body and detached from the possession of any finite entity, the "I" increasingly devoted to becoming unlimited faces a bottomless precipice: "should I jump without a rope or protection towards the Infinite Consciousness' arms?". The fall, now, is a free fall, without control or a guarantee of returning. Final fall that also creates an interesting impasse: is it better to remain secure, wise, and fluid in *vijinánamaya kos'a* or to throw myself into the abyss of the plane of "*hiranyamaya kos'a*", or literally, "the golden layer?" The agony of such doubt is like a storm. Even though the "I" is very, very close to the divine point of "Self" and can practically grasp it, it is not possible for the "I" to touch it. Without facing the abyssal sink that separates the "I's" dot from the point of the "Self," without facing the risk of completely surrendering oneself, without going forward with the suicidal and renascent jump into nothingness, there is no way to advance. It is finally time to face the much-despised question and find out: after all, "who am I?"

P. R. Sarkar (2013) will say that "there is a 'self' inside every living being" (p. 36). It is in this "sense of 'self' of every individual" (idem) that the essence of spirituality is hidden. This "self" who now writes, who wanders in social media, who sits down to meditate. In a nutshell: this "self" who is connected to external objects and, by connecting to them, feels that "I exist." This is *mahat*, the mind's most subtle portion. No longer the "I did" of the mental screen of *citta*, or the egoic "I do" of *aham*. Simply "I exist." I know by intuition that, somehow I feel that I exist. But where do I exist? How? My Facebook profile says that I exist in images that tell me that I am Camila Mozzini-Alister, a PhD in Communication at UERJ and a PhD in Artes: Producción e Investigación at UPV, born on 06/08/1987, married to Ajit Mozzini-Alister. My mantra says: "I am Infinite Consciousness." "I am Infinite Consciousness." "I am Infinite Consciousness." "I am one with It." "I am one with It." "I am one with It." "I am one with It." Yes, whether it be through the existence of my profile, whether it be through the existence of my voice internally uttering who I am, I feel that I know I exist. And it is in this feeling of "I know the fact that there is a sentiment of 'I exist' in me" (idem) that resides the *átman*: "the 'self' of 'I know,' the subject of the sentence "I know that I exist,' is *átman*, the Spirit, the Soul" (idem). The only thing left is to take a chance: to assume the risk of knowing that I exist not only by the act of accessing my Facebook profile, but also to question myself if, in fact, I exist as Infinite Consciousness. The Infinite Consciousness is the place, the final point with which I must merge my dotified "I" and put an end to this much-tortuous question. Without fear: I jump.

"I am Infinite Consciousness." "I am Infinite Consciousness." "I am Infinite Consciousness." "I am one with It." "I am one with It." "I am one with It." "I am one with It." From the jump that drowns the "I" in the Cosmic Ocean's depths comes the ascension. Enchanted by the *mantra's* vibration, the latent divinity of the *kúlakundálini*, the serpent coiled around *muládhára cakra* wakes up. It wakes up and awakes the four *vrttis* of desire: material desire, psychic desire, psychic-spiritual desire, and spiritual desire. Drunk with desire, *kúlakundálini* then marches toward the *svádhisthána cakra*, right above the *muládhára*. After getting there and balancing the hormones secreted by the gonads and the six petals of despise, stupor, indulgence, lack of confidence, cruelty, and the idea of annihilation, the "self" takes pleasure in *sálokya samádhi*. As the first stage of "*samádhi*," or "union with God," in *sálokya*, the "self" "feels that, on the level which one finds themselves, in the sphere where one's exulting mind is, they are not alone" (Ánandamúrti, 2013, p. 61). The pleasure enjoyed by this feeling of being always accompanied by the Supreme allows the coiled serpent to keep uncoiling itself. By uncoiling, it not only goes up, but when it traverses the third *cakra*—*manipúra cakra*—it reestablishes the hormonal secretions of the adrenal glands and the pancreas and its ten affections materialized into the *vrttis* of sadism, shyness, envy, lethargy, melancholy, irritability, thirst for desires, attachment, anger, and fear. Here, the dot of the "I" experiences *sámipya samádhi*. "*Sámipya*" means "proximity." Proximity that takes pleasure in the feeling of "I got really, really close to the Supreme Being, I am very close to the Supreme Progenitor. Thanks to my spiritual practice, the distance between It and me is being reduced" (Ánandamúrti, 2013, p. 62).

"I am Infinite Consciousness." "I am Infinite Consciousness." "I am Infinite Consciousness." "I am one with It." "I am one with It." "I am one with It." "I am one with It." By facing the satisfaction of having persisted in the meditation and attuned all the vibrations of the energetic vortexes located in the three lower *cakras*, *kúlakundálini* is elevated toward the *anáhata cakra*, the fourth *cakra*, which is located at the heart's level. Now, a dozen more affections closely related to the glandular secretions of the thymus are soothed: hope, worry, effort, love, discernment, mental paralysis due to fear, vanity, ego, avarice, hypocrisy, propensity to arguing, and regret. And within this soothing, the "I" experiences a third stage of *samádhi*, the *sáyuja samádhi*, which means "in intimate contact with, side by side, already touching it" (Ánandamúrti, 2013, p. 62). If in the first stage, *sálokya*, the "self" feels that "God is with me," in the second, the "I" feels Its proximity, "in *sáyuja*, one feels the tactile experience" (idem). Delighting itself in the pleasure of touching God, the serpent of the dormant deity moves to the throat and penetrates the *vishuddha cakra*, the thyroid gland, and the manifestation of its energetic vortexes related to the sixteen sonorous expressions: peacock sounds, ox sounds, goat sounds, deer sounds, bird sounds, donkey sounds, elephant sounds, acoustic root of creation, preservation and destruction, the sound of *kúlakundálini's* elevation, the sound of the transmutation of theory into practice, expression of mundane knowledge, expression of well-being in the most subtle sphere, expression of acting with nobility, expression of giving oneself to the Supreme, repulsive expression and, finally, sweet expression. At this point, the "I's" dot experiences a more subtle *samádhi*—the "*sárupya samádhi*"—whose feeling is not only one of intimate contact, but also of communion: "I am one with It," "I am one with the Supreme Progenitor, with the Supreme Cognition" (Ánandamúrti, 2013, pp. 62–63).

However, the vibrating dot of the "I" wants to be even closer, more tangible, more united with the immense magnetism exerted by the Supreme point of the "Self." How to resist this intense attraction? "I am Infinite Consciousness." "I am Infinite Consciousness." "I am Infinite

Consciousness." "I am one with It." "I am one with It." "I am one with It." "I am one with It." Attracted by the pleasure of experiencing the feeling of uniting with the divine, *kúlakundálini* wants more: now it ascends to the *ajina cakra*, the third eye between the eyebrows, and engages the secretions of the pituitary gland, balances the *vrttis* of the mundane and the spiritual knowledge and takes pleasure in *sársthi samádhi*. Exulting, overcome by the feeling of expansion that invades it, the unitary dot of the "I" and the Supreme dot of the "Self" are tuned to the feeling of "I am It," "I am It." "I" and "It," two subjects united by the connective link of the verb to be. Two entities and one verb. Even though they are coinciding, there is yet another distance to be abolished: the complete union, the final stage of *samádhi* admits the existence of just one subject, one entity connected to the verb to be. And tipsy with desire for a total and complete fusion, for the "I's" absolute expansion beyond space and time, the slumbering serpent wakes up and climbs to the mountain's summit until it touches the pineal gland and reaches *sahasrára cakra*, the *cakra* of a thousand petals. At last, the meeting of points takes place. Bathed in light, electrocuted by golden rays, the shiniest gold glitters makes the body tremble in the divine soul enlightened by the verb "to be." Among the ecstasy of a joy that cannot be found in any masturbation, orgy, or social network orgasm, the "I" then melts away into tears because it has finally reached the *átman*, the soul, the place beyond the mind where one just "exists" without any differentiation between "I" and "Self." "I am." "I am." "I am." "I am." "I am." "I am." "I am": I am Infinite Consciousness (Fig. 3.10).

***

There I was, in July of 2014, sitting in lotus position, smiling and crying at the same time because I was feeling limitlessness. No longer bound to the past, present, or future. No longer restricted by territory, geography, or body. Beyond space and time, without beginning, middle, or end: subtle, as subtle as possible to an "I." So subtle that I uncoiled the turns of the "I's" circle, shattered its roundness in loose dots, submitting it to the funnel of discipline until the only surviving point of this journey vibrated in the mantra to reach the subtlety of the divine point. The straight line that separated them has succumbed. And the vibrating dot melted into light, a golden light that elevated the slumbering serpent of *kúlakundálini* and spun the circle of the *cakras* at the same time that it spun the wheel of

Fig. 3.10 Samádhi and the union between "I" and soul, by the author

samsara, the cycle of life and death, by the burning of *samskaras*. Electrocuted by the tantric shock, the "I" is enlightened in ecstasy. With the body cleansed from any kind of dense energy, with the eyes, ears, nose, mouth, sense of touch, hands and feet, sex, and excretions becoming introverted, with everything that was attached to the outside, to the exterior, to the other, turned off. With the fingers firmly intertwined on my lap, they no longer type or glide on planes of smartphone devices. The device became my own mind. And the planes to be touched, now, are the seven planes of the mind through the harmonization of the fifty affections contained in the seven points of *cakras* located along the spine. Untouchable, the proximity of the divine touch has increasingly separated the "I" from the human touch. A touch that brought the "I" back to the body, to the flesh, to the ground, to the living room, to space and time. A touch that broke the trance and, in the unpleasant longing for that *samádhi*, turned the search for meditation into an obsession.

References

Ánandamitrá, A. (1999). *Beyond The Superconscious Mind*. Ananda Marga Publications.

Ánandamúrti, S. (1979). *Ánanda VacanÁmrtam Part 12*. Ananda Marga Publications.

Ánandamúrti, S. (2001). *Ananda Marga Ideology and Way of Life in a Nutshell—Part 5*. Ananda Marga Publications.

Ánandamúrti, S. (2005). *Psicología del Yoga*. Ananda Marga Publicaciones.

Ánandamúrti, S. (2008). *A Liberação da Mente através do Tantra Yoga*. Ananda Marga.

Ánandamúrti, S. (2013). *Ánanda Vacanámrtam—Parte 1*. Brasília: Ananda Marga.

Bernardo, Gustavo. (2011). Um espelho retorcido. In: FLUSSER, Vilém; BEC, Louis. Vampyroteuthis infernalis. São Paulo: Annablume.

Singer, B. (2010). Modernidade, hiperestímulo e início do sensacionalismo popular. In L. Charney & R. Schwarz Vanessa (Orgs.), *O Cinema e a invenção da vida moderna*. Cosac Naify.

CHAPTER 4

Medi(t)ation and the Desire for Omnipresence

The contemporary technologies of communication almost fulfil the dream of omnipresence, which is the dream of making the lived experience become independent from the places that our bodies occupy in space.
Hans Ulrich Gumbrecht

Since July of 2014, my everyday habits started to be guided by the rhythm of these two anthropotechnics of the biological substratum's expansion: inner expansion and outer expansion (Fig. 4.1). In the morning, the first thing to be done right after waking up involved the superhuman effort to avoid getting caught in the morning laziness by conditioning the body in the strict discipline of meditation practices: drink a cup of lukewarm water with lime[1], perform *ásanas* to control my lower *cakras*, dance the *kiirtan*[2] mantra and then sit down to meditate for at least twenty minutes while

[1] An ancient habit of the Ayurvedic medicine that seems to be turning into a new fad in the editorials of healthy life style magazines, the daily consumption of lukewarm water with lime juice alkalinizes the blood by performing a thorough cleansing in the body's acidity and toxicity levels.

[2] *Kiirtan* is a dance with specific movements created to balance the pineal gland and prepare the mind for meditation thanks to the repetition of the Sanskrit mantra "Baba Nam Kevalam," which can be translated as "love is all there is." A little demonstration of the mantra is available in this video: https://www.youtube.com/watch?v=-3MPJoVsXPs, accessed on 07/27/2020.

C. Mozzini-Alister, *Does Social Media Have Limits?*,
https://doi.org/10.1007/978-3-030-74120-4_4

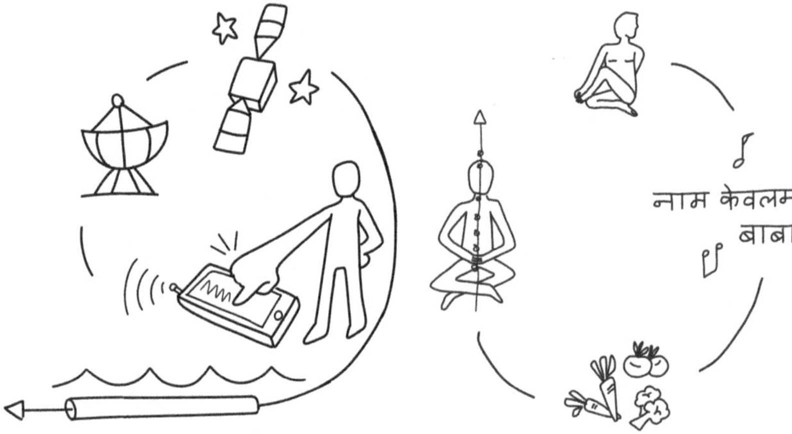

Fig. 4.1 The outer expansion via mediation and inner expansion via meditation, by the author

reciting the respective mantras of opening and closing, and the *Guru Puja* to offer the thoughts I had throughout the process of meditation. With each new practice, my body stopped being just a body, my "I" was no longer Camila Mozzini-Alister and started to become a single dot in the cosmic space enchanted by the vibration of its mantra. With each new attempt at deactivating the senses, the mental burning of actions and reactions accumulated in *samskaras* would set the body on fire, which would speed up my experience of living. With each new effort to concentrate the mind, there would be an exponential growth in the expectancy of being possessed yet again by the strength of that golden light that made me infinite. And with each new meditation attempt, the frustration of once again not having achieved such state would grow as well: what aspect of me would have to be improved? What part of my affections was blocking the direct touch with the divine point of the "Self"?

Finished that intense period of introversion, it was time for the much-expected moment of extroversion through the undisciplined practice of mediation: as if in an impetus to be rewarded for the failure of my previous endeavor, the thirsty fingers would run to deactivate the smartphone's airplane mode, log in the Facebook App, and frantically spin the mind through the newsfeed's scrolling. To alleviate the seriousness involved in the meditation process by entertaining myself with the spinning of other people's thoughts was like a well-merited dessert after such a mental hard work. With each new login, my body stopped being just a body, my "I" stopped being Camila Mozzini-Alister in the geographical space and started to become a planetary point covered by dots in the topological space of the network's social validation. With each new update of the mantra "what's on your mind, Camila?", my own mind would become indexed as an action of a profile always accessible to the mere touch of fingers belonging to any given hand. With each new post, emoji, comment, and share, the expectation that the simultaneity between action and reaction would once again deflagrate the euphoric sensation of embracing the world of "Us/ Knots"[3] of Facebook profiles would grow stronger. And with each new frustrated attempt, that agony of, again, not being able to materialize the proximity of the touch from those who were no longer by my side would become even more critical: what was I lacking so that I could feel the touch of distance? What blockade of my affections was insufficiently attractive to associate far away and close by with the proximity of immediate relationships?

Marooned by this double frustration of the desire to touch, to leave the device's body behind and go back to being only a body without a double required an even more superhuman effort than the already superhuman effort to meditate. Before such difficulty, it was not rare for activities such as cooking, eating, doing the dishes, and doing the laundry, going out, getting back home, and walking around, going to classes, research groups, and artistic meetings, concentrating on lectures, chores, and writing, to become increasingly converted into activities performed in parallel with the digital connection to my double. The touchscreen cell phone had to always be ready, the closest as possible to my hands so that, in the event that the mental alarm of connection would go off, at any given moment I could immediately look at it and check what was going on in the reality of the profile dot linked to all the other dots. Thus, synchronically to the morning conditioning of meditating, I also developed the intensive conditioning to, without no quarter whatsoever, mediate. And after finishing my daily tasks in tandem with the act of continuously accessing Facebook, at night another round of ***kiirtan***, mantras, and meditation would take place. Dividing myself in order to become one with everything and uniting myself again to end any kind of division, the act of living became schizophrenically marred by a double contradiction: secretly offering my

[3] The Us/Knots concept was previously explained.

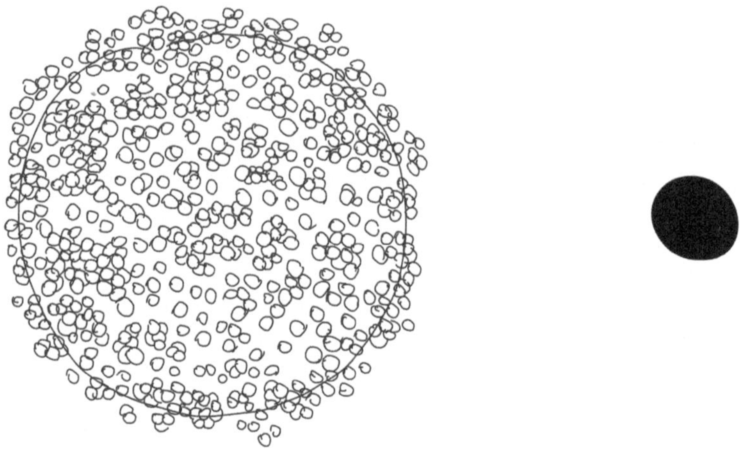

Fig. 4.2 The self as the Great Mind of Facebook and as the Infinite Consciousness, by the author

unexpressed thoughts to the Cosmic Mind and extravagantly scattering others around as "thinking-actions" in the unceasing mixture of my own self with the Great Mind of "Us/Knots" of Facebook (Fig. 4.2).

Between the lonely introversion of mental images and the collective extroversion of posted images, my life compulsively spun like a dog chasing after its own tail: when meditating, it became a micro-point in the cosmic space, and when mediating it was transformed into a macro-point full of micro-dots connected in the topological space. Meditating, mediating, meditating, mediating, meditating, mediating, meditating, mediating, meditating, mediating, meditating: mediating and meditating and mediating. With each new dawn, I was immediately assailed by the anticipation of feeling once again limitlessness. With each new dusk, I would feel once again the frustration of this unfulfilled expectancy. Expectation and frustration: the customary interaction with this cyclical pair of affections made each new morning another day to exercise the faith to overcome them both. An heroic overcoming, winning by exhausting these two enemies that would insist, day after day, in coming back to nourish the even tastier hope of victory. "Vengeance is a dish best served cold," as the well-known saying goes. A vengeance that, by combining the hunger with the willing to eat, that is, the obsession of becoming infinite and the addiction of feeling connected, culminated in the creation of an explosive cocktail: the merging between the duo mediation and meditation into the practice of a medi(t)ation that linked the desire for the mind's enlightenment to the pleasure of the electrocution of fingers on the screen. In the urge to expand every limit that would confine the "I" to the space occupied by the body, the joint search for enlightenment and electrocution culminated in the development of techniques and control mechanisms applied to the manipulation of an extracorporeal anatomy. What kind of anatomic control is this that makes us go beyond limbs, torso and head, muscles, bones and tendons, arteries, veins and ducts, organs, glands, and systems?

Moving away from the traditional medical compendiums, what I was interested in on a daily basis was the double exercise of a subtle anatomy. The first, by incarnating in the profile through the shock of fingers on the screen to incorporate the device's senses, making myself globally present via wireless waves that circulate by cables, antennas, and satellites; the second, by subtilizing the mind's seven screens in order to awake the *kúlakundálini* and in *samádhi* make me present in all spaces and times through waves that circulate in the *vrttis* of the seven *cakras*. No longer the mechanical circulation of legs, wheels, engines, and wings, but of forces that, whether it be by touching points on a screen, or by touching points on *cakras*, capture the biological substratum's electromagnetisms in favor of extracorporeal modes of circulation. Such urge for electrocution and enlightenment, in order to be present in a space outside space and in a timeless time, ended up subsiding that which I call "desire for omnipresence." A desire that is put into action by the continuous sophistication of the approximation modes of the multicolor light of Facebook profiles, and the golden shine of *Parama Purusa*, arising from a systematic control of the forms of energetic circulation: a "General Economy of Light"[4] that carries out the project of transforming the body into a body of light.

Etymologically, the word "economy" comes from the merging between the Greek root words "*oikos*" (home) and "*nomos*" (law) and relates to the territory of one's home and all forms of management pertaining to its

[4] For Phillipe Dubois (2004), photography "is in any case a curious matter of light, or rather, of circulation of light" (221), which implies the need to "rethink all photography in the context of a general economy of light, which concerns not only photography, but also cinema, video and painting" (idem). Taking into account the possible repercussions of this questioning, we will try to study here a "general economy of light" from the circulation circuits of the affections modulated by the production of images in the mediation and meditation binomial.

inner space. However, "home" is no longer merely the place where we live and management is no longer restricted to quantities capable of being objectified and measured: we dwell in the roundness of our own bodies as well as in the topology of the profile dot. And the management of these strongholds of interaction to oneself and to others is not related only to the control of the circulation mechanisms of tangible elements such as individuals, consumer goods, and populations. When introjected the political-economic mechanisms of state regulation of populations[5] (Foucault, 2008) and embedded the biopolitical strategies[6] that make "life

[5] The ascension of a political economy is related to the formation of government States, which no longer concern a territoriality but a populational throng that is both the instrument and target of management. Spread out across Europe from the sixteenth century onwards, points out Foucault (2008), security devices are consolidated as technologies of power in the eighteenth century, a period that is directly linked to the emergence of an art of governing, whose core principle considered the population as a problem. That is the case because, beyond the demographic expansion, the money abundance, and the increase of agriculture production, the conditions of possibility that cemented an art of governing in States of government are related to the problematic of populations. Thus, population emerges, henceforth, as a political and technical object of government. It is at this moment that the notion of economy as the management of the family shifts, configuring a political economy that no longer has the family nucleus as its end and goal. Family becomes, then, from a model to a privileged element and instrument of government among the population. The problem now is how to organize optimal ways to circulate the populations, such as shifts, exchanges, contacts, and modes of dispersion and distribution that are useful to the State. What matters, in this context, is no longer the imposition of laws, but the tactical disposition of things based on a liberty that does not want to allow anything and everything, but that understands that a certain level of permissiveness is useful and even indispensable. Hence, while sovereign and disciplined societies were organized through interdictions of movement that put every individual in their own place, in societies of security a "certain" movement is stimulated since it would allow for an ever-increasing level of action and reaction, uniting freedom and circulation.

[6] Foucault (2008) links the notion of biopolitics to the third and last configuration of the market proposed by the American Neoliberalism or Anarcho-Liberalism. More than a political and economic option, Anarcho-Liberalism "is a way of thinking and being. It is a kind of relationship between the ones who govern and the ones being governed, going beyond a mere technique of the ones governing towards the ones being governed" (p. 301). In it, we have a crystallization of the enterprise model as a social organizer based on the expansion of market processes toward territories considered as non-economic: the journey of the individuals during their lives and the investments that they received and created generate, then, a human capital to be produced, accumulated, and paid for according to the competencies of each person. Inequality and competition built the biopolitical foundations of individuals who manage themselves as if they were a business. It is up to every individual to know the risks that one is willing to take and every person needs to be responsible for their own safety. A

and its mechanisms enter the domain of explicit calculations" (Foucault, 2010, p. 155), what is at play is the acquisition of another type of capital. No longer a monetary or human capital: what matters now is obtaining what I call "energetic capital," the only one sufficiently abstract to transform the human experience of a body into a superhumanly omnipresent body. By facing the wide array of available possibilities to hoard and control the individual and the collective energy levels, mediation and meditation currently appear on the horizon as very promising anthropotechnical methods in the process of creating the "human as *ex-tasis* of its animal condition" (Romandini, 2012, p. 9), that is: the flesh and its instincts.

Nevertheless, the complete consummation of this ambitious project of overcoming the body does not happen overnight. Just like a lamp, a star, or any other luminous entity, in order for a body to constitute itself as a body of light, first and foremost it has to be a source of energy: there is no street light without electricity, in the same way that there is no Sun without a nucleus, and there is no smartphone without a battery. Thus, the first dimension of intervention in the contemporary "General Economy of Light" (GEL)—making a connection to the concept of "general economy"[7] proposed by Bataille (1987)—is to occupy ourselves with "the movements of energy on Earth"[8] (Bataille, 1987, p. 48). Or more precisely: to study the processes that produce the electrocuted/enlightened body as an energized body due to the maximization of its electromagnetic

society of autonomous beings. Unemployment and differences in wealth would no longer be problems to be solved, but necessary elements to establish inequality as society's structural condition and to invigorate the game of competition. Due to this fact, by biopolitics, we understand that which "makes life and its mechanisms to enter the realm of explicit calculations, and transforms the knowledge-power into an agent of transformation of human life" (Foucault, 2010, p. 155), operating an acute potentialization of life based on a useful distribution of the living that gives more and more "access to the body" (idem).

[7] In "The Accursed Share," contrary to the conception of a "particular economy" that does not deal with spending for the sake of spending, useless, Bataille (1987) proposed the idea of a "general economy" as a transversal field of studies whose problem can be thought "based on any discipline that deals with the movement and the energy on Earth" (p. 48). Such as the Sun, that shines and provides a superabundance of the necessary energy for life without charging or demanding anything in return, Bataille posits the existence of an economy of exuberance. Wealth, in this sense, emerges no longer as a retention of things, but as the promotion of energy circulation. Thus, the key point, contrary to the accumulation of scarce goods or the satisfaction of necessities, is now the luxury, poetic, and invigorating spending of the energy surplus produced by the Sun.

[8] The author's own translation from the Spanish: "*desde cualquier disciplina que se ocupa del movimento de la energía en la Tierra*".

potential. Here sketched based on the anthropotechnical encounter between mediation and meditation, such potentialization will be outlined by the sequencing of three lines: a vertical line, a horizontal line, and the closing of such line in a circle. In the first, we will tackle the process of energetic elevation by the polarization of affections; in the second, the approximation between screen and mind through the neutralization of affections; and finally, the achievement of the mind enlightenment and the body's electrocution by closing the energetic circuit. In each one of these stages, three variables are at play: body, screen, and energy (B/S/E) for mediation; and body, mind, and energy (B/M/E) for meditation. However, being both material substrata to the divine touch, screen and mind coincide in their condition of medium: from the Latin "*medius*" and from the Indo-European root word "*medhi,*" a common root word associated with the terms "meditation" and "mediation" that is related to the intermedium space, what is between, pertaining to a channel, a duct, a passage that is never complete in itself because it can only be a halfway. Mediums that, in their complementary opposites, meet in order to intermediate the encounter between two other variables: body and energy.

Let us begin with the first stage that concerns the process of energetic elevation by the polarization of the body's less elevated affections. We can see above us a vertical line, related to the neighboring space outlined by screen and mind (Fig. 4.3). Neighboring because it is located between two other extremes: the energetic condensation of the matter-as-body and the complete rarefaction of the matter-as-energy. Body and energy are vertically separated, away from each other by a distance that takes place in strata, levels, and longitudinal gradations: one has to go up. In this sense, the first intervention to suppress such distance is to inscribe a point into the line, a mark that can instruct a target to be pursued with the vectorial help of an arrow: from the bottom-up, from the body toward the apex of its energy. One must then subtilize the body. In other words, make it less dense and less populated by wave frequencies that create mental images bearing "bad vibes," incapable of being shared on the App's screen where the profile dwells. After all, it is not sensible to purposely expose to other network dots moments of intense suffering, pain, scourge, sadness, anguish, fear, and envy. And even though sometimes a few exceptions will happen—such as the famous video of singer Sinead O'Connor[9] and episodes of suicides

[9] Complete video available at: https://www.youtube.com/watch?v=919ip9_urEo, accessed on 07/27/2020.

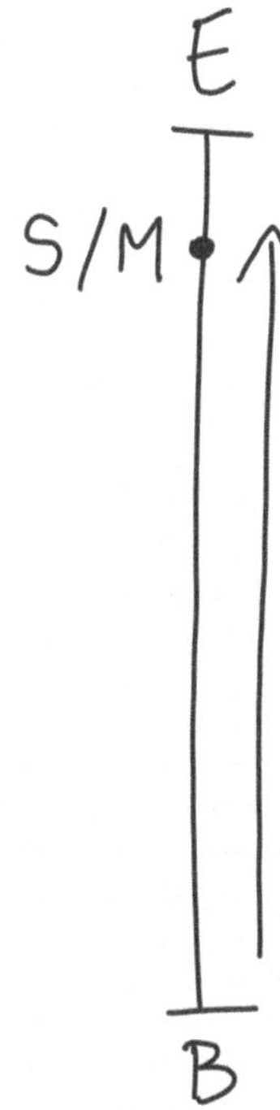

Fig. 4.3 The straight line of energetic elevation by the polarization of affections, by the author

broadcasted live—we hardly feel encouraged to post texts, videos, or pictures of the intense crisis of despair, crying, mourning, insecurity, anger, panic, or anxiety related to the affections of the "lower *cakras*."

As the name already indicates, lower *cakras* are the ones located below, the lesser or "non-higher" ones because they store the most undesirable remnants of human animality: the so-called survival instincts, that is, innate patterns genetically inherited from ancestral lineages of *Homo sapiens* that have the goal of perpetuating our species. From the Latin "*instinctus,*" its meaning is related to something that stimulates, internally instigates and constitutes a type of intelligence in its most primitive state: the one responsible for securing the self-preservation and the collective reproduction through the four basic instincts—hunger, sleep, sex, and fear. We are helplessly programmed to eat, sleep, have sex, and sense possible risks by means of the *vrttis* of the first three *cakras*: the material, psychic, psychic-spiritual, and spiritual desires of the *muládhára cakra;* the warnings of despair, stupor, indulgence, lack of confidence, cruelty, and the idea of annihilation of the *svádhisthána cakra*; and the defensive strategies of shyness, sadism, envy, lethargy, melancholy, irritability, thirst for desires, attachment, anger, and fear inscribed into the *manipúra cakra*. It is precisely this programming of the biological substratum directly related to the mind's first three layers—*annamaya kos'a*, *kamamaya kos'a*, and *manomaya kos'a*—that we are being summoned to deprogram in the contemporary General Economy of Light.

How to contain the desire to be such as we project ourselves in our best images? Traversed by this search, currently allied with happiness imperatives that promise the existence of states of absolute and immediate gratification via consumption (Rocha, 2007), we replicate such logic through the happy publicizing of ourselves on a social media where the possibility of "disliking" does not exist. We might like, love, smile, get surprised, share, comment, or even get angry or sad, but we can only feel angry or sad for such events, and not disapprove them. If a beloved relative dies or a romantic relationship ends, there is no way to "dislike" it. Thumbs up, signaling that "everything is alright," "we agree," "things are good," "all positive," is not only the iconic symbol of Facebook but it also crystalizes a subjective "wave of liking" by the positive programming of the affections. And governed to externalize, on screens, increasingly elevated facets of ourselves in order to acquire the energetic capital of the others' attention on the trail of a mere like, the monster hidden in the basement of the human mind becomes more and more inflated. Inflating and vibrating in images full of non-positive affections. What can we do with all the rest that is thought but it is not liked or expressed because it is interdicted in the "negative" category? Conscious or not, some will spit their own monstrosity on the very topology of the network or even vomit themselves in the geographical space so that they can deflagrate and then receive the immediate reaction of their actions. But there is also a way to control them by using *ásanas,* diets, mantras, and by the continuous rhythm of the *kiirtan* that directly impact the energetic circuits of the *cakras.*

In this scenario, the focus of control is then redirected toward the control of the effort by means of the body's constant conditioning to the adherence to habits focused on the mind's progressive subtilization. If previously the body would wake up and scroll its laziness on the bed, drink coffee, do some physical exercise or immediately leave to perform some work-related activity, such routine is now completely modified. Before working, it is necessary for the body to function like a machine in order to deprogram its programmed routine of habits: to wake up and already make an effort to avoid stalling; before even eating, to exercise in yoga positions aimed specifically at realigning the *vrttis* of the lower *cakras*; to persist in such harmonization of the affections by uttering mantras in Sanskrit and by redesigning once and for all the neural connections and the hormonal secretions that affect the nervous system. If previously, the alarm clock would go off half an hour before the daily toil would begin, now said toil starts at least one hour and a half before any activity is performed. Sometimes, even before the Sun has risen. And when leaving home, the training does not stop: any signs of a possible negativity, whether it be in the traffic or when arriving at spaces of social interaction, one already sets off the immediate mental activation of the mantra; if there is a social event that involves eating or drinking, foods such as alcohol, meat, mushrooms, onion, and garlic are avoided and, usually, will provoke personal conflicts due to the pre-judgment of peers pertaining to what one decides to put inside one's body. Lengthy questionings and the oddest possible inquiries become a constant and have to be daily addressed. It is not rare that such inquisition is accompanied by a changing in the circle of friendships: sometimes, social interaction would be restricted to groups that would only reinforce these new habits; other times, old groups would be preserved while new ones are added, thus creating a schizophrenic salad of energetic environments. The important thing is to persist in the continuous practice of transforming self-discipline into a way of life to be shared in the great circle of dots that cohabit in the social network space.

Thus, together with the disciplined effort to control the patterns of what is thought in the mind, also comes the continuous effort to control what others will think about the "thinking-action" exposed on the screen by the profile. To the control of the habits concerning the elevation of the mind, another one is added: the habit of constantly electrocuting the fingers on the device's skin to better exert control over our self-image. After all, how can we be better seen in what is projected by our double? The first step is to interact. To be constantly posting, commenting, sharing, and activating emojis to not only appear in the newsfeed highlights, while feeding off its reactions, but also to distribute them in the scrolling of other "thinking-actions." As the old saying goes: "out of sight, out of mind." And since after a few days everything that is posted falls into a collective oblivion and helplessly becomes just another file on the profile's personal page, there is no other option but to post again and again and again, indefinitely. It is in this continuous incitement to update ourselves through the simple movement of fingers that, by osmosis, we also learn a reinforcement code of our own image as a positive self-image. Motivational messages, personal accounts, smiling pictures, curiosities, erudite quotations, animal videos, vacation pictures, existential questions, frustrations, news, conflicts. The effort to reinforce how good, engaged, cool, intellectualized, spiritualized, funny, and pondered we are goes hand in hand with the expectation of an immediate agreement, acceptance, and reinforcement of the effort of what is positively projected of oneself. Unintentionally, "marketing" subtlety inhabits the tinniest corners of subjectivity (Deleuze, 2008) and assigns the self-promotion as part of the most sophisticated codes of mental behavior.

Done. Due to the arrow's guidance at last we have achieved the desired landmark: with the mind's images and the images of the screen being properly elevated into a programmed grammar of affections, the energetic point's summit is finally reached. The "self" is now "on top" of its negative affections, electromagnetically edited by the control of the *cakras'* energetic circuits as well as by the control of the device's energetic circuits. By being made completely positive, "what comes from below can no longer affect" one's enviable superiority. And as if losing one of the poles due to the intense climbing toward the affections' summit, the vertical line also loses its negative polarity and, from its foundations, falls into the most perfect horizontality. A smooth fall, without any noise or impact. The climbing to the mountaintop of the affections transformed the steep relief into a smooth and slippery open lowland in which, on one side we have, again, the energetic condensation of matter-as-a-body and, on the other, the complete rarefaction of matter-as-energy. Body and energy are still separated, but now by a distance that takes place on a latitudinal scale that cannot be measured in kilometers or measurement units capable of being calculated: in it, one searches for a proximity to the soul as well as for a proximity to whoever is no longer physically by one's side. Is it crazy to want to touch what can only be felt by the sense of smell? Against this scenario that appears to be impossible, there is no other option but to take a risk. To take a risk and, with arms wide open, proceed to bet on the oscillations of the free movement through the spaces of the mind and the screen scattered on the horizon of this straight line with one single hope: to find the point of balance between body and energy.

In this new plea, the main ingredient is: to feverishly persist. Persist in the mechanisms of control of the energetic circuits of the mind and the screen so that, when they are expanded, they can mutually balance each other out. Thereof constitutes the effort of overcoming: to give the best of oneself in *ásanas* even when one is exhausted; to eat subtlety even when one wants to eat junk; to dance *kiirtan* even when one wishes to perform an uncontrolled move; and, twice a day, hide away to meditate even when one wants to be outside, in the world. For that, the effort of meditation must be an even more reinforced effort: to advance in the realms of the mind, one must take literally the statement that "we must keep our mental balance and be indifferent to both compliment and slander, and to happiness as well as to sadness" (Ánandamúrti, 2013, p. 58). And when the possibility of the negative is excluded from the horizon, the word of order now is neutrality, after all, positivity became a synonym for impartiality. Neither too much nor too little, neither more nor less, neither above nor below, neither cold nor hot: always lukewarm. Do not laugh out too loudly and do not cry too much. Do not hug too much and do not be too distant. Do not expose yourself in excess and do not stay too quiet. Do not take sides but also do not stop being outraged. Apolitically, the affections become "flat," always serene and anesthetized by the continuous deprogramming of our basic instincts. And such constancy never entails less meditation: to achieve a state that is immune to every kind of external disturbance is only possible through an even more concentrated pulsation of the mantra during each cycle of inhaling and exhaling air from our lungs—and preferably, for more than twenty minutes. But it is at this point that, mild and moderated, by oxygenating the cells and spreading the vital energy of the *pranah* across the nervous system, the mind progressively expands itself beyond the space occupied by the body: it increases the possibility of enlightenment because the body's neural connections become more energized and stable.

However, the mind's point is not yet a point of equilibrium. It is increasingly stable, shielded from the vibration of images that disturbs it and it is closer and closer to the Cosmic Mind—source and stronghold of every form of energy—but miles and miles away from the body. One must then compensate such disparity by bringing the screen closer to the body, the hands, the ears, the skin and, at last: the fingertips. If by meditating, the mind receives more and more energy and expands itself because it suspends and introverts the body's sensory-motor organs, by mediating it does the exact opposite. By increasingly exciting and externalizing the senses and locomotor limbs simultaneously to the constant shock of fingers on the screen, more and more energy will be received by the device tuned into the Cosmic Mind of Facebook. A mind whose energy is offered to us through the pleasure of electromagnetically touching images outlined in the scrolling, checking, posting, tagging, using emojis, commenting, and sharing. Thus, as if it was another body cell, the cellphone becomes a vital part of the daily caresses and a faithful companion of the day-to-day life: at the gym, at work, at the classroom, at the library, at the table, at a bar, at the restaurant, at the concert, at the park, on the beach, in the countryside, in the car, on the train, on the airplane, at the waiting room, on the boat, at the airport, in the bed, in the living room, in the kitchen, in the backyard, in the bathroom, in the purse, in the pocket, in the ocean, at the pool, at the protest, in the carnival party, at the mountaintop, at the bike ride, on a hike, on a hang glider, on a morning run, during the gentle pace of a baby stroller, at the wedding, at the church, at the spiritual retreat, at the funeral, at childbirth. And also standing still: from where one has never left and has not even thought about the possibility of physically moving.

With a few exceptions, practically there is no place left where the body can go that the screen cannot follow behind. Touchscreen, always ready and willing to obey the commands of our fingers and open the App to receive, and to witness on the network, the best thing we can offer of ourselves through the social sharing of a video, an image, a text, a link. Counterbalancing the lack of social interaction that arises from the extension of periods of affective neutralization to further expand the mind, the practice of mediation is a balm to soothe the curtailing of the human contact due to the ever-increasing levels of interaction with people's profiles. Quick and practical, without the need to waste time on the act of moving the body to spaces of interaction. Now, the persistency no longer takes place due to the rigidity of the effort, but also through the unruly entertainment that arises from the constant act of updating oneself and others in the immediate reverberation of the mantra "What's on your mind?". To that end, appearing "interesting" to others becomes the norm: when facing the large scope of comparison that spins across the newsfeed cogwheels, the strategies to catch other people's attention become even more sophisticated. There are those who will opt for the "long post" and will dedicate themselves to write entire thesis to discuss issues they deeply care about and that are, frequently, controversial. But the fact is that, it does not matter if the text is long or short, with or without a photo or a video, or full or empty of emojis: to positively express emotions, even the most "negative" ones, requires true mastery. Mastery because it no longer matters if the published format's content is happy or sad, festive or mournful, angry or placid, engaged or resigned, plaintive or grateful, scandalous or moderated, atheist or spiritualized: if posted, it has to be edited in a way to be approved by the selective pre-requisites of the self's energetic capitalization as a profile. As a good self-advertiser, one quickly learns that only by attracting other people's attention to interact that the energetic quotient stored on the device, through the electrification of fingers on a screen, will at last obtain a quantifiable feedback. Quantifiable no longer monetarily, but in the form of numbered notifications that, by measuring the degree of instantaneity of the reactions to the profile's *thinking-actions*, create the feeling of embodiment of the presence of those who are no longer by one's side (Fig. 4.4).

With an accelerated breathing and the heart beating faster, the euphoria produced in this feeling of concreteness experienced by the energetic proximity with the other makes, this time, the screen expand in the opposite direction to the expansion of the mind: toward the excited body whose nervous system is electrified through the act of touching images. However, stirred by the screen, the mind needs to go back to meditating in order to again become pacified. Nevertheless, after having its peace restored, the quiet mind longs for the body and goes back to the screen to socially interact and get excited again. Like a serpent that eats its own tail, the cyclic duo does not stop: expansion of the screen, expansion of the mind, expansion of the screen, expansion of the mind, expansion of the screen, expansion of the mind, inner expansion, outer expansion. When the vectorial single-pointed arrow of the affections' elevation falls on the horizon of the straight line, it divides itself and doubly wanders in a parallel way from here to there, from there to here. Meditating and displacing the mind to the energy's corner, mediating and moving the screen to the corner of the body. In this daily flux of consecutive expansions, some effects are crystallized. The point that would previously oscillate from one side to another, first displacing the screen toward the body, then displacing the mind toward the energy's corner, slowly becomes stabilized toward the center of the straight line until it finally reaches a middle ground. Since it

Fig. 4.4 The energetic oscillation between mind and screen, by the author

followed all the protocols requested by the medi(t)ation methods, the point of equilibrium located at the perfect equidistance between body and energy brings forth its reward: the rendezvous between mind and screen. From this collision, however, another dimension of the biological substratum's subtilization also arises: the training of the mind in the exemplary detachment from the touch and the entertaining self-editing of oneself in the touch of the screen, ally themselves into a single program. By vehemently desiring to be just like the spiritual master and to share each step toward such achievement with everyone as a profile on the network, the "I" turns itself not only into a model, but above all into a "desirable model." In other words, an available matrix to the energetic consumption of the modes of living as much as the image of gurus, famous profiles (or the so-called influencers) and sneaky announcements on the newsfeed disguised as posts (Fig. 4.5).

Etymologically, the word "model" comes from the Italian "*modello*," a diminutive of "modus," which means "way," "measure." A measure that in its proper adjustment becomes a mold to be mimicked and reproduced through force or example until it can consolidate itself as the norm. From that arise the mode, the fashion, the framing, and all the variety of derived meanings that, even though distinct, have a common concept: there is a certain ideal prototype capable of being abstracted, analyzed, and internalized beyond its simple and practical applications. Thus, after having found the exemplary point of equilibrium in which body and energy are symmetrically canceled by mutually complementing themselves, the meeting between mind and screen kick-starts a new stage of transition: the propagation of the measure between mind and screen beyond the space of the line. Such intent is only possible by keeping oneself even more devotedly focused on the just middle ground: on the unmoved peace of the body's social isolation, which is compensated by the inexorable interaction of the mind via profiles on the screen. It is at this median point, detached from the concrete but attached to the concreteness of the image, that through the effort of persistence, mind and screen dilate their extension in order to encompass the complete amplitude of the straight line. "Extension," a

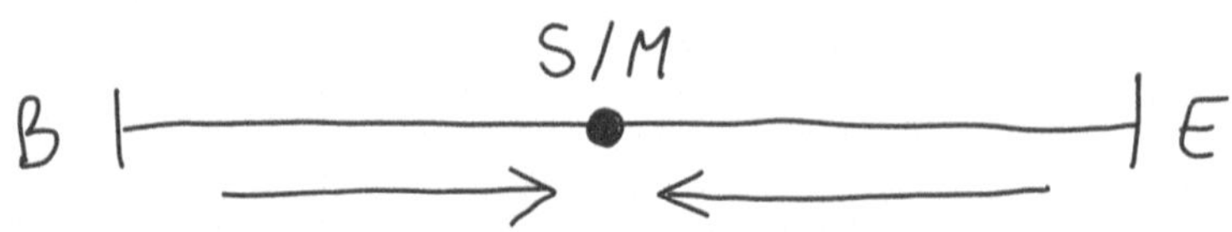

Fig. 4.5 The middle point between mind and screen, by the author

word from the Latin "*extendere*," composed by the merging of the prefix "*ex*" (toward the outside) to the root word "*tendere*" (to stretch): with the goal of becoming divine and extending one's divinity beyond space and time, the screen is extended to the maximum proximity to the body and the mind is completely distended toward the energy vortex. And from the communion between both expansions—the mind's inner expansion and the screen's outer expansion—arises an association so powerful that it not only laterally encompasses the entire possible extension of the straight line but it also forces the inversion of the straight line's vectors toward the outside. By feeling more and more "different," more conscious and separated from the amorphous rabble through the skillful energetic reprogramming of the affections, the "I" self-legitimizes itself in its own mental advertisement as a cause and a measure of everything that needs to be followed, liked, and shared (Fig. 4.6).

It only takes a few minutes of scrolling through Facebook to realize the propagation of causes pertaining to the "I" that seeks the support of others: feminist and sexist, homophobic and LGBTQIA+, left-wing and right-wing, Republican and Democrat, environmental and market-friendly, vegan and carnivore, meritocratic and social driven, spiritualist and atheist. All kinds of polarities. There are those who would explore the nuances that exist in these extremes with the aim of creating a dialectical synthesis between both; there are those who would "like" something here and comment on something there but, without taking sides, they limit themselves to only replicate videos, images, and texts with curious and unusual situations. What matters is that each person may not only follow others but also be followed by other people. The only requirement is to learn how to build a niche of attention so that everybody wants to be there: on the newsfeed shouting and following the latest events that take place in this virtual forum of discussion updated in real time. It is only "natural" that this would happen not only because social networks are interesting by themselves. Allied to the fascination created by mediation, there is also the

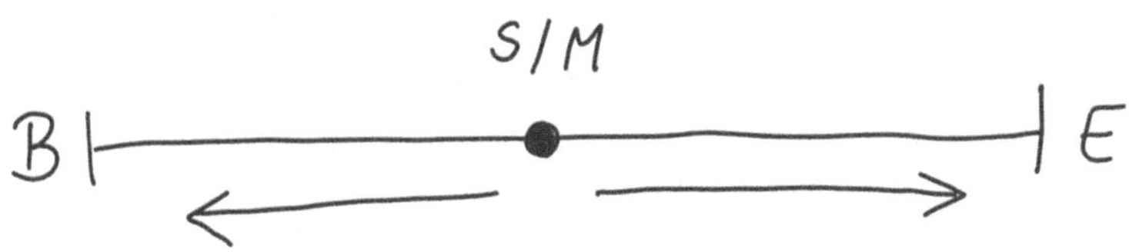

Fig. 4.6 The mental advertising of oneself as a model, by the author

success of meditation routines in tenaciously maintaining the lower *cakras* under control: any desire but the spiritual desire is gradually interpreted as a devolution. And since one is only allowed to evolve, little by little the detachment from the physical touch is intensified and transformed into a detachment from the pleasures of the flesh, from the mundane. And the field of battle, is turned into an eminently mental arena. This is why we must connect the mind to the profile's mind on the screen, the only one capable of performing the immediate outpouring of the energetic revolution that anticipates every single day into a new day of a "UFC" game: let the fight begin! Even more conscious, more militant, and more empowered, the "I" then takes on the noble role of spokesperson and expands the mind and the screen to the limits of what is thinkable: to the boundaries between body and energy. However, by bumping into the fact that it can no longer expand, it does not resign itself, kick starting a new form of progression: the unconscious alteration of the perception of reality through the twisting of the extremities of the once straight line into a curve (Fig. 4.7).

What is now at stake is the very questioning of the statute of the real: "how can one say that the mental reality of the profile does not produce a body if, within said reality, we not only daily become affected by it but also fight on it?" Due to the conviction that the field of battle was displaced to the topological ring, the "I" struggles and no longer suffers to change the concreteness of the world, but above all to change the models through which the world is perceived (Flusser, 1985). The revolution then becomes "the" revolution of information, knowledge, of sharing of knowledge, of the YouTuber who bends the straight line's rigidity. And by investing even

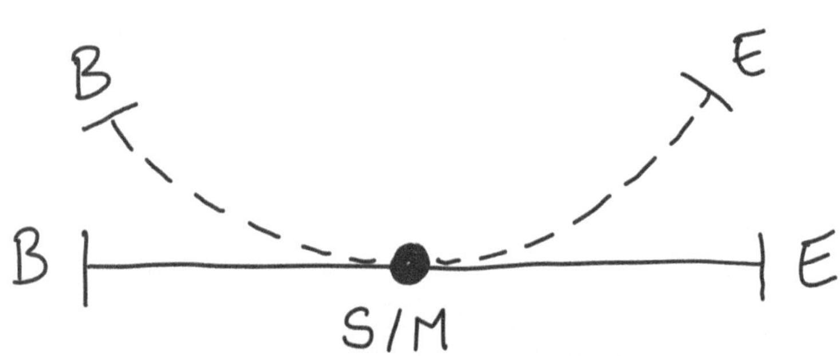

Fig. 4.7 The change in the perception of reality due to the curvature of the straight line under the model of the social media activist, by the author

more time and vital energy—*pranah*—into the advocacy for one's own causes through the mental change of the other materialized as a profile on the screen, the first symptoms of deception arise. As hard as one tries, argues, informs, and comments, the "less woke" other helplessly ends up being a lost cause. Bounded to the brutality of ignorance, these others not only refuse the information but also search for counter-information in fake sources even less informed than them (Silverman, 2015). Disillusioned, the "I" with a doctoral degree in the rulers that measure the just and the right slowly becomes detached from all of those who are strange to them: they block, exclude, unfollow, and reduce even more the diversity of "Us/Knots" in the already reduced network that does not mingle with the gentile who drinks, smokes, eats meat, does not meditate, and does not believe in anything. To avoid contaminating their body with any kind of negative energy source, the square root of social detachment culminates in indifference—a necessary quality to transform the previously engaged "I" into a divine judge. And from the highest pedestal of sobriety, it utters its sentences. It decides what is good and evil about everything it considers to be outside its own mold by distilling posts filled with sarcasm, ironically commenting on other people's posts, indirectly taunting profiles and so on and so forth... Until it reaches the highest degree of moderation: the criticism of the criticism through the acidic dissemination of posts against the very existence of Facebook.

The contradiction is that, even though the "I" is exhausted with the stupidity of others who do not listen to its commandments and, quite conscious of the advertisement game played by the codes of programming and behavior of said social network, it still does not quit it. There it stays, cynically disdaining the despicable inferiority of the human race as if it belonged to any other species but that one. It stays because, when all the dimensions of desire that compose the first *cakra's vrttis* are spiritualized, the judge gives up legislating and starts searching for the last available orgasm: the mental orgasm. Since its basic instincts were stunted, satiating the need for food, sleep, danger prevention, and sex become mere mechanical activities performed so that the body can operate in the physical world only for the sake of the mind. No longer savoring the pleasures of a greasy food, of a long and lazy nap, of the heat of a good fight, or of the carnal sex so that it can simply meditate, the mind hungry for spiritual pleasure is fulfilled on the screen. Compensating the desire for food in order to be feed by information through reading and seeing articles, links, images, and videos located on the newsfeed's scrolling. Counterbalancing the time

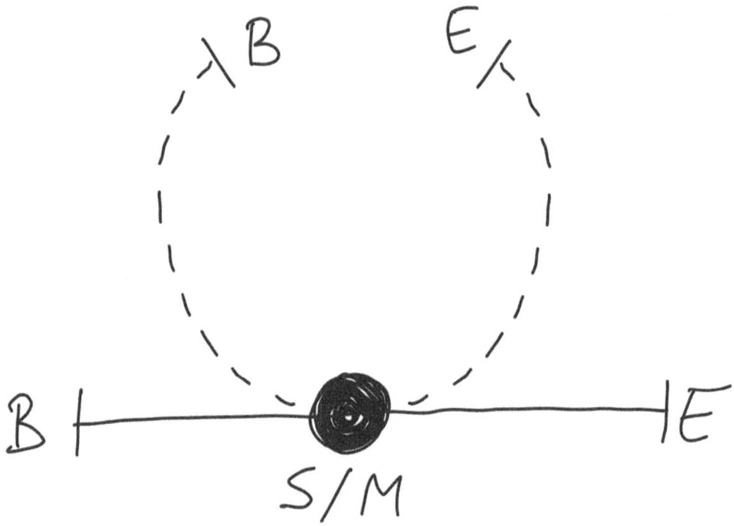

Fig. 4.8 The voyeur's indifference curvature, by the author

wasted on sleeping so that it can always be online, always active in the Messenger's chat and its conversations. Neutralizing the desire to curse and send the most disgusting beings to hell and beyond in order to wisely talk to a selective group of carefully chosen followers. And at last, sublimating the will to, like a wild animal, fuck, for the unpunished and innocent masturbation of profile images on the screen. Before the personal revolution of such discovery, the edges of the straight line begin to bend even more strongly since the judge ignores its rulers and promotes itself for having achieved the last degree of indifference: the apathy of the *voyeur* (Fig. 4.8).

"*Voyeur,*" from the French "*voir*" and from the Latin "*videre*" (to see), is the one who gets sexually excited by the secret contemplation of other people's intimacy. Without ever consummating the carnality of the touch, the traditional *voyeur* willingly takes pleasure in the placid observation of scenes of live explicit sex as well as in similar scenes that take place at specialized movie theaters, in pornographic movies or in online pornography websites that currently constitute one of Internet's greatest allures. Nevertheless, the medi(t)ative voyeurism is more subtle. First, because generally, the persons in question hardly know that they are acting like a *voyeur*. Second, because in the innocence of never looking for a nude body, the mind unconsciously compensates this forbidden debauchery through the energetic bond of the shock between the fingers and the screen that takes place via profile images. And so it goes. With the tongue, the nose, and the genitals all hungry, the *voyeur* excites the vision, the sense of hearing, and the tip of the fingers in the scrolling of the newsfeed, from profile to profile, devouring pictures, savoring albums, smelling posts: fiercely taking possession of the intimacy of others as if it would dominate the very body it looks at through the screen's servile obedience. Overwhelmed by the obsession of literally wanting to translate the yogi maxim that the body is nothing more than energy, the *voyeur* becomes excited in the feeling that it secretly exchanges caresses with people even when simply touching profiles. Secretly because, unless it leaves a digital trail, the identity of a voyeur is never revealed: always accessible to each and every one, there is no way of knowing who puts the fingers to better see the profile's parts unless a like, a comment, a message, or a share are materialized. Nonetheless, this is what constitutes a good *voyeur*: the apathetical ability of sordidly looking without leaving any trace of what was seen. It is not by chance that so much energy is invested in capturing the attention of others by advertising oneself to its best model: it is not enough wanting to be desired, there is also the desire to discover how, when, where, and who truly desires us without having the courage to explicitly do so.

Always busy doing nothing, the practice of procrastination is introjected as a part of daily habits: hours and hours of life are daily wasted so that one can enjoy the mental subtleties of this divine touch because medi(t)atively subtilized. Subtilized by the maximum abstraction of the brutal condensation of solid matter and its wet liquids, hormones, and sweat responsible for exciting the primitive senses of smell and taste. Without the saliva from a kiss, without the viscosity of the mucus, without the skin's sweat, without the vulva's lubrication, and without the sperm from the testicles. Dried, practical, and without risk, socially incited and individually yearned, the divine touch between bodies that collide through the touch of profiles on the screen starts a sublime orgy: the electromagnetic orgy of mental waves transmuted into radiofrequency waves that come and go as image dots accessible to each and every finger that is tuned to a smart device on Facebook. The *voyeur* game is a blind game and, because of that, a quite pleasant one: its modulation conflagrates the model of the complete actualization of oneself and of all the other selves as a mutual panopticon (Foucault, 1987) that is all-seeing, all-knowing, and that potentially interacts with the other who is always within sighting reach. This is what became largely known as the "Bentham Panopticon," which was idealized in 1795 by Jeremy Bentham as an architectonical model applicable to the prison system: from above, in a central tower, a single watcher could observe all the inmates without them knowing that they were being watched. In the uncertainty of the watcher's presence, the mere fear that someone could be there would be enough for the prisoners to adopt the behavior desired by the watcher. However, it is in the virtual showcase of dots present in the Great Mind of Facebook that the panoptical model is subjectively amplified: it no longer needs towers, walls, and bars, and the fear is transformed into the fear of being invisible and, therefore, undesirable in the immense web of profile dots that composes the network. It is no accident that the "I" seduces so well that it obscenely insinuates itself to others with bikini photos, sexy poses, and provocative angles. It does so in order to be seen in its maximum visibility so that, by decentralizing the single panopticon observer to encompass everybody, everyone can now globally participate in the voyeur game both as the *voyeur* and the "*voyeurized*."

Hence, it is in the reciprocal offering of oneself to the unrestricted touch of the other's gaze by uniting mind and screen that the "I" becomes closer to God thanks to the desire of becoming closer to everyone: "I am Infinite Consciousness," "I am Infinite Consciousness," "I am Infinite Consciousness," "I am Infinite Consciousness," "I am one with It." Actualizing the meaning of its personal mantra via the hyper-connected existence of the *voyeur*, the mind enlarged to the size of the world relishes, but also feels guilty. And by feeling guilty, it remembers the austerity of the yogi's body to once again become centered. Marooned by the fact of being everywhere at the same time, the "I" then begins the protocol of introversion of all *indriyas*—the ten sensory-motor organs—and sits down to meditate: the hands are fixedly interlaced on the lap, the legs are crossed in lotus position also immobilizing the feet, genitals, and excretory organs, the vocal cords become mute, the eyes close, the tongue folds on the palatal dome, the sense of touch is shut off from the physical contact with the screen, and the ears and the nose are protected in an environment free of any smells or sounds. Completely motionless and in command of each one of the voluntarily controlled organs, the "I" concentrates on the mental purification of the images so that they become crystalline: without the sexual excitation of the *voyeur*, without the desire of eating images, without smelling profiles, or without moving around through all devices. However, while it gradually shuts itself off from external stimuli and immobilizes the desires of *kamamaya kos'a* already quite subtilized, the subconscious mind (*manomaya kos'a*) feels a deep attraction to the device's body: "If only we didn't need to eat. If only we didn't desire to fuck" (Acconci, 2001, p. 374) as the castrated and gutless screen can quite well do... Overwhelmed by this unconscious "nostalgia not so much about the past but as a fiction of the future" (idem), the "I," by wanting to turn its own body (*annamaya kos'a*) into the body of the device, puts into question its own sense of existence. After all, how do I know that I am truly real if I can no longer perceive myself in the human form? (Fig. 4.9)

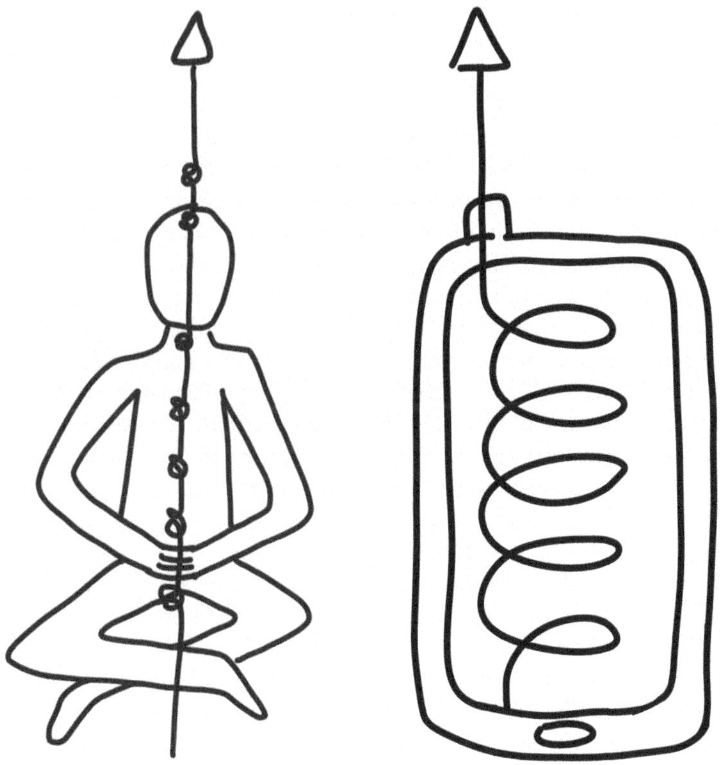

Fig. 4.9 The inverted mirror between yogic body and apparatus body, by the author

Absolutely assured to become free from all basic instincts, the meditation deepens. All thoughts then vibrate to the sound of the *Ista mantra* so that one's sense of identity can become completely dissociated from any remnant image of the body. By becoming a self-voyeur, the "I" observes its existence only through the light of the image of an Infinite Consciousness: "I am Infinite Consciousness," "I am Infinite Consciousness," "I am Infinite Consciousness," "I am one with It". Nevertheless, facing the frustration of the ever-repeated *samádhi*, the mind once again searches for the multicolor existence of the profile in order to rediscover the infinitude that once it felt. Image of the image, a double of itself turned into a dot by the biological substratum's disintegration into pixels, bits, and electronic impulses capable of being irradiated via radiofrequency waves. Within the timid exposition of the soul tangential as a personal profile page by means of the energetic connection between fingers and App, the "I" finds a shortcut to the channel that allows the point to "cut through the radius" (Ánandamúrti, 1979). That is the reason why, right after the meditation comes to an end, the "I" once again grabs the body of the device—more important than any other person—to get entertained. Entertained by the burning of *samskaras* through the karmic acceleration of the interactive interplay between actions and reactions in the social media's life: it states, comments, shares, partakes in other people's *samskaras*, and creates new potential reactions until the point when all the fun ends. The "self" becomes irritated. However, having already learned how to ignore everything that comes from the lower *cakras*, the brain synapses' affective control is quickly restored into a constant state of mental positivity. And in its unshakeable positivity, the "I" propagates itself by turning that simple entertainment into something serious: it committedly advocates, indifferently legislates, and ironically criticizes until it remembers, in the apathy of the *voyeur*, the ultimate sense of its existence: playing God. A Super-Human God who knows It exists not only because It can see, know, and reach everything through the simple shock of fingers on a screen, but also because It is recognized and socially worshiped as an image and a likeness of Itself on the network.

From now on, whether it be mediating or meditating, in neither of those cases the "I" no longer observes its existence happening in the shape of a body: it is either the image of the Infinite Consciousness of *Parama Purusa* or it is the image of the image of itself in the profile's Infinite Consciousness that holds the world in the palm of the screen. Thus, even more isolated because openly co-opted by both the social network's algorithmic filter bubble[10] and by the spiritual bubble's sectarian filter, the detachment from the body is converted into an inexorable attachment to the realm of the image. By fantasizing inhabiting the mind's hallucinogenic reality that doubts the doubt of its own physical existence (Flusser, 2011), the "I" finally ends up completely bending the straight line until it becomes "a magic circle around society" (Flusser, 2008, p. 33). It is only in the circularity of the closed circuit that the limits that previously separated body and energy once and for all become united through the blind conjunction between mind and screen. And no longer seeing any kind of boundary or duality between the body's energetic condensation and the corporeal rarefaction of the energy, henceforth there is only an infinite loop: electrocuting on the screen in order to enlighten the mind; enlightening as a profile in order to electrocute in *samádhi*. Lost in the bedazzlement of a structure that only ends at the beginning, the "I" daydreams in the imagination of having achieved the dream of divine perfection: the perfection of the formless image that can never become deformed. In other words, the "I" falls prey to the illusion that it is no longer impacted by the affections of the physical body inherent to the experience of space and time due to the delirium of being omnipresent, omniscient, and omnipotent in every "here" and in every "now." Thus, no longer accepting the existence of any form of distance, the Beyond-Human God quickly moves toward a single option: the complete convergence of the "Western super-device in the form of a mass culture programmed by increasingly better-coordinated devices" (Flusser, 2008, p. 79) and the "Eastern super-device in the form of the rigidity and expansion of its own device" (idem) until the existence of the body is finally actualized into a body of light (Fig. 4.10).

In search of, at any cost, materializing the divine form in which the "I" so well observes as its real sense of existence, the "I" sinks even deeper into

[10] Coined by Eli Pariser in the book "*The Filter Bubble: what the internet is hiding from you*," launched in May 2011, the notion of filter bubble is related to situations in which the algorithm of a website selectively "guesses" what information the user would like to see based on the data generated by the own user, such as the exposition of tastes, locations, or the history of clicks. Pariser was worried about the algorithms becoming so good at delivering exactly who and what we interested in that they can start limiting our openness to new ideas and people, creating types of culture bubbles that would isolate us from other possible realities.

Fig. 4.10 Closing the circle and completing the process of transforming the body into a body of light, by the author

the precipice of the image. And without a thorough evaluation of the consequences of its creations, it automatically adheres to the entertaining use of wearables such as googles,[11] gloves,[12] and sound and voice command

[11] Even though the investment in virtual reality googles does not materialize the ultimate end of the anthropotechnical endeavor, companies like Facebook have not given up on them.

[12] An example of such technology is Tilt Brush, something akin to a new version of "Microsoft Paint for the year 2020"—says its developers, the technological innovation company Fast Company—which was created from a series of vision and manual manipulation devices that announce the possibility of "Painting tridimensional strokes in real-life size, stars, light, or even fire. Experience painting like never before." More details at: http://www.tiltbrush.com/, accessed on 07/27/2020.

devices[13] that alter its perception of reality's molecular structure: the "I" no longer interacts with an environment of things, but with an environment of rays. And taking the idea that "magnetic and galvanic phenomena of nature" (Zielinski, 2006, p. 26) materialize "the presence of God in the world" (idem) to its last consequences, the Beyond-Human God takes on the heroic role of energizing everything, no matter at what expense. Exploiting even more the environment and the "lesser humans" in order to inhabit the holographic reality of the social media's profile that knows everything, sees everything, and is capable of everything, the Supreme Engineer follows its journey toward the unbeatable novelty. After assimilating the knowledge that the shock of its fingers can resuscitate the screen's death, it decides to play as the animator of all objects. Refrigerators, television sets, alarm clocks, airplanes, clothes irons, pens, wristwatches, cars, and the most banal items[14] become as smart as the screen of the smartphone.

Fascinated by its creation, it sees that all of this is good. However, the power to give life to objects is still not enough to satisfy the essence of its existence and It wants more. After connecting more and more utensils with the firm conviction of turning the world into a limitless place after its own image, it moves toward a new experiment: to telepathically connect minds and objects. Why should we waste time pressing buttons, plugs, or even brushing our fingers on smooth photosensitive screens if thinking is already enough? And the news of this latest goal are announced across the four corners of the Earth. From the highest echelons of the Zuckerberg

[13] An example of this, besides the famous voice software known as "Siri" (for the iPhone), is the voice assistant created by Google, called "Google Home." Launched in May of 2016, it has the shape of an elliptic speaker from where one can control the entertainment system across one's home, it helps organize daily errands, and it can also be used to access Google's search engine. More information at: https://en.wikipedia.org/wiki/Google_Home, accessed on 07/27/2020.

[14] Better known as the "Internet of Things" (IoT), the phenomenon pertains to the interconnection via the Internet of computational devices embedded in daily objects, allowing them to send and receive data. Based on the growing use of the so-called RFIDs (*radio-frequency identification devices*), Hayles (2009) proposes the production of a new ontological dimension of the relationship between humans and objects that is no longer based on the binarism and the hierarchy of the human. More information at: http://www.zdnet.com/article/what-is-the-internet-of-things-everything-you-need-to-know-about-the-iot-right-now/, accessed on 07/27/2020.

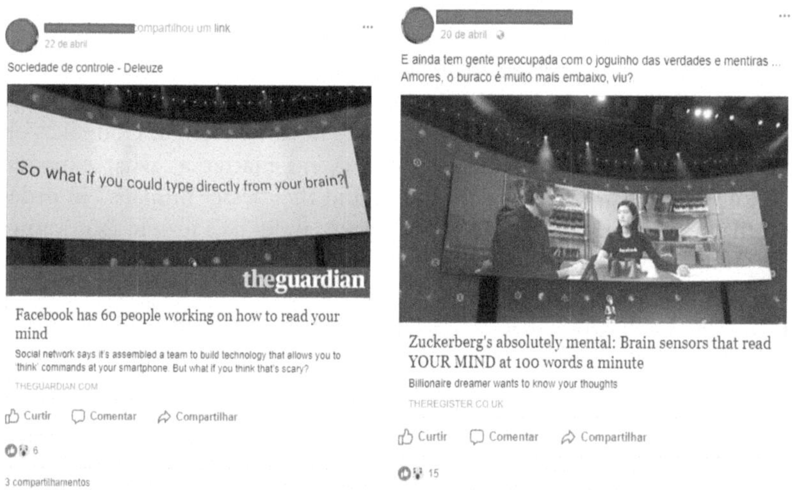

Fig. 4.11 Posts collected on mind-reading technology developed by Facebook, from Facebook App

Realm[15], brain-computer interface engineers, neuroscientist and neural image engineers are diligently working on developing a system capable of typing 100 words per minute straight from the source: the brain (Fig. 4.11).

Diving even deeper into the screen's depths, the "I" decides that it is time to eliminate the screen itself. From now on, in order to be real, everything must become as ethereal as the mind, and reality is now "augmented reality." Before the temptation of accelerating even more the time of sharing in social media, the "I" resolutely accepts to completely format its thought as mere machine-like processes. All of this in order to become more and more quickly deified. And bedazzled with its own geniality, it now also dominates the decoding of brain waves into convertible signals to the binary code of bits. Bragging about its ever more subtle subtilization, it realizes that all of this is good. It is so good that it does not require any surgical implants or electrodes: merely through the use of non-invasive

[15] Complete article on *The Guardian* newspaper can be found at: https://www.theguardian.com/technology/2017/apr/19/facebook-mind-reading-technology-f8, accessed on 07/27/2020.

sensors, the All-Mighty Beyond-Human finds out that it can do more. The immortal Shakespeare already said: "we are such stuff as dreams are made on." Dreaming about the consummation of a reality more real than reality itself, the "I" now aims at discovering the telepathic reality of dreams. From that point on, minds will be connected to minds. Objects are connected to objects. Minds are connected to objects. And objects will be connected to minds. From high above its masqueraded puritanism that at any cost refuses to feel the pleasures of the flesh, It can only conceive Itself as an Angel. And since Angels have wings, all kinds of motion sensors are created so that it can fly: eyes, limbs, facial recognition, thought, voice, environment. When one lives in the Magical Universe where everything is unlimitedly connected, everything also becomes equally intelligent. And when everything becomes indistinctively smart, the project of affective reprogramming carried out by the current General Economy of Light reaches its final destination: the majestic model of the resurrected body of Christ (Romandini, 2012).

Angel *par excellence* and the outcome of divine intervention's first biotechnology (Romandini, 2012), it is in the subtleness of the Resurrected Christ that the Beyond-Human Cherub at last finds the form of its existence. A form that traverses every wall and encounters no boundaries, moves elsewhere and faces no delays, soars high above and overcomes even gravity itself. But above all, a form capable of "perceiving without being affected" (Romandini, 2012, p. 217). Impassible by having lucidly achieved a non-animal sensitivity and, therefore, Beyond-Human, the spores of such conquest only timidly appear to the present: increasingly intertwining the energetic circuits between mind and screen in the slowly but surely approximation between mediation and meditation. Medi(t)ation, two programs of expansion of the biological substratum, united in favor of the desire for inhabiting the profile's mental reality under the texture of the hologram: the Substance of the Spirit. It is in the consistency of this tridimensional image produced by a coherent source of rays that, by completely dissociating the body from the mind, the Angelical Realm of the Image becomes the ultimate reality because it is free from any duality. With no inside or outside, without real or virtual, with no online or offline, without person or profile. In this limitless reality, all the meanings of "unification" presented in the word "yoga" are fulfilled in its maximum connotation. Everything is connected. Everything is One. Everything is infinite. Diluting the screen and capitalizing the mind into holograms that no longer feel hunger, fear, the urge to sleep or have sex, the Earthly Angel finally becomes like the profile in the body of its adored smartphone: the resurrected body of Christ. Therefore, disincarnating from its own body to rise as a body of light, the "Divine Self" takes pride in having found the meaning of its existence while forgetting about one tiny detail: in order to resurrect, we first have to die.

References

Acconci, V. (2001). Art & Authority. In G. Moure (Ed.), *Vito Acconci—Writings, Works, Projects*. Barcelona: Ediciones Polígrafa.

Ánandamúrti, S. S. (1979). *Ánanda VacanÁmrtam Part 12*. Fiesch, Switzerland: Ananda Marga Publications.

Ánandamúrti, S. S. (2013). *A Graça do Senhor—Parte 2*. Brasília: Ananda Marga.

Bataille, G. (1987). *La Parte Maldita*. Barcelona: Editorial ICARIA.

Deleuze, Gilles. (2008). *Conversações*. Tradução Peter Pál Pelbart. 7 ed. São Paulo: Ed. 34.

Dubois, Philippe. (2004). *O ato fotográfico e outros ensaios*. 8 ed. Marina Appenzeller (trad.). Campinas: Papirus.

Flusser, V. (2008). *O Universo das Imagens Técnicas—Elogio da Superficialidade*. São Paulo: Annablume.

Flusser, V. (2011). *A dúvida*. São Paulo: Annablume.

Foucault, Michel (1987). *Vigiar e Punir*: nascimento da prisão. Trad. Lígia M. Ponde Vassalo. Petrópolis: Vozes.

Foucault, Michel (2008). *Nascimento da biopolítica*. Tradução Eduardo Brandão. São Paulo: Martins Fontes.

Foucault, Michel. (2010). *A História da Sexualidade II*—O uso dos prazeres. Trad. Maria Thereza da Costa Albuquerque e J. A. Guilhon Albuquerque. 20th ed. Rio de Janeiro: Graal.

Hayles, K. (2009). RFID: Human Agency and Meaning in Information-Intensive Environments. *Theory, Culture and Society, 26*, 2–3.

Pariser, E. (2011). *The Filter Bubble: What the Internet Is Hiding from You*. London: Penguin.

Rocha, S. P. V. (2007). *Yo no creo em brujas, pero que las hay, las hay*: a publicidade, o consumo e a crença descafeinada. *Comunicação, Mídia e Consumo. São Paulo, 4*(9), 127–137.

Romandini, Fabián Ludueña (2012). *A comunidade dos espectros*. I. Antropotecnia. Alexandre Nodari e Leonardo D'Ávila de Oliveira (trads.). Desterro, Florianópolis: Cultura e Barbárie.

Silverman, C. (2015). Lies, Damn Lies, and Viral Content: How News Websites Spread (and Debunk) Online Rumors, Unverified Claims and Misinformation. *Tow Center for Digital Journalism, 168*(4), 134–140.

Zielinski, S. (2006). *Siegfried Zielinski: genealogías, visión, escucha y comunicación*. Bogotá: Ediciones Uniandes.

CHAPTER 5

Limits

> A Klee painting named "Angelus Novus" shows an angel looking as though he is about to move away from something he is fixedly contemplating. His eyes are staring, his mouth is open, his wings are spread. This is how one pictures the angel of history. His face is turned toward the past. Where we perceive a chain of events, he sees one single catastrophe which keeps piling wreckage upon wreckage and hurls it in front of his feet. The angel would like to stay, awaken the dead, and make whole what has been smashed. But a storm is blowing from Paradise; it has got caught in his wings with such violence that the angel can no longer close them. This storm irresistibly propels him into the future to which his back is turned, while the pile of debris before him grows skyward. This storm is what we call progress.
>
> Walter Benjamin

We are so well carrying out the anthropotechnical project of suppressing the basics instincts of survival—hunger, sleep, sex, and fear—that we do not realize we are on the verge of extinguishing ourselves and performing the suicide of our species in the name what we call "progress" (Fig. 5.1). By wanting to be something we are not and never will be, Angels, we are distorting our desire for expansion in order to become increasingly alienated from our own humanity. "Alienation," from the Latin "*alienus,*" which means "from the outside," "belonging to the other." And more and more out of synch with reality, unfamiliar with the social events and productive chains of the neoliberal economic model, like

C. Mozzini-Alister, *Does Social Media Have Limits?*,
https://doi.org/10.1007/978-3-030-74120-4_5

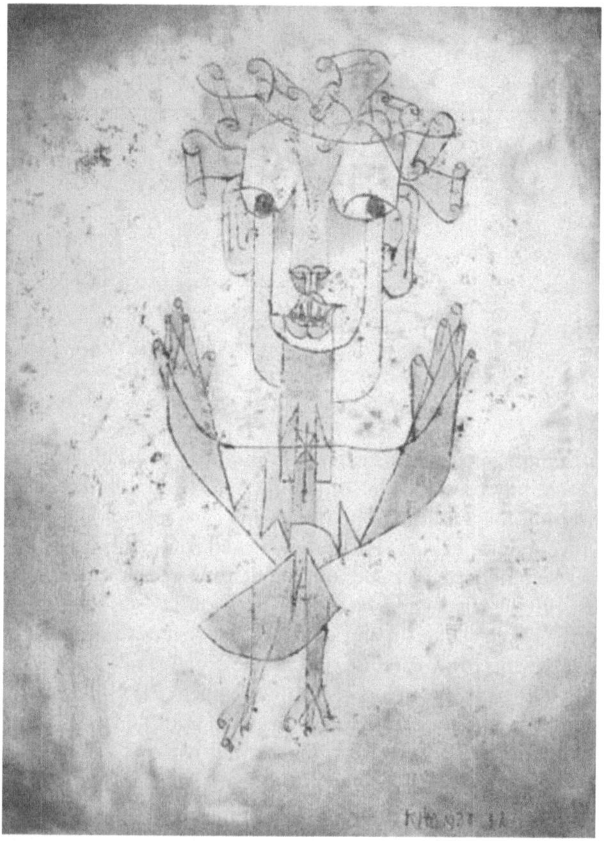

Fig. 5.1 "Angelus Novus," by Paul Klee

"aliens" among our own species, we no longer see the true cost of our fascination for smartphones: the enslavement of African children that precariously work on the extraction of the minerals required for the production of smart devices[1], the mass suicide of Chinese employees working at "telecom" companies[2] that look more like terrible slavery manors, and the

[1] Complete article at: https://www.youtube.com/watch?v=WysjWqQ0oc8, accessed on 07/27/2020.

[2] Complete article at: https://www.theguardian.com/technology/2017/jun/18/foxconn-life-death-forbidden-city-longhua-suicide-apple-iphone-brian-merchant-one-device-extract, accessed on 07/27/2020.

brutal devastation of natural resources caused by human action in the middle of the "Anthropocene Era." That is, of a historical period in which human intervention is so intense that changes and acts like a biological force of global impact upon the climate and the management of our planet's ecosystems.

Our devices smell like blood. As Benjamin said so well, "there has never been a monument of culture, which is not simultaneously one of barbarism" (Benjamin, 1994, p. 225). Nevertheless just like well-behaved alienated people, we firmly believe that we are at the highest point of evolution, at the apex of progress. Since the advent of the press, then the telegraph and the telephone, the mail and the telegram, the radio and the television, the most utopian dreams of social transformation are conceived as technical miracles carried out by communication devices. The radio would be the "redemption of the illiterate," television would make "the end of disinformation" possible. As Vaz points out (2010), "in history, the experience of a technological opening is usually followed by a feeling of absolute singularity" (p. 192). And infused with a blind faith that the present is the highest of evolution, we naturalize our technologies as if they were a logical result of progress and not what they truly are: just a manifestation, a material product of the subconscious mind (*manomaya kos'a*) of this human animal desperate to know itself.

Who knows by heart where the insignificant vesicle is located, its function, and what vital substances it secrets prior to fatally losing it one day? Ignoring the basic knowledge of the body's materiality, we indirectly seek to know that by exteriorizing it on the so-called Materiality of Media. First, by emulating the ears through the advent of the radio and technical sound devices, after that, emulating the eyes with the invention and the dissemination of cameras, and now, emulating the sense of touch via the banalization of screens photosensitive to the touch. Flusser would say, "we create instruments to simulate any aspect of our existence, and these instruments counter-attack, and then we simulate these instruments and know ourselves better" (Flusser, 2015, p. 99). We already manipulate the sense of hearing, the vision, and the sense of touch, and up to this point all is well, everything is alright, thumbs up: all easy. However, the rabbit hole goes deeper, "further down," when one thinks about the two rawest senses: the sense of taste and the sense of smell. Directly related to the liquid and the solid elements of the energetic condensation of matter, there is currently no way to be a living being without inconveniently having to eat, drink, and deal with the secretions and excretions of everything one eats and drinks. To end this once and for all, we adopt the most radical of solutions: we become images.

We put on virtual reality googles, gloves, and motion, touch, and sound sensors, we exploit even more the environment and the "less humans" so that we can inhabit the social media profile's holographical reality that knows, reaches, and sees everything. While the body eats food infused with chemicals, synthetic dye, preservatives, and highly toxic meat that comes from companies that profit from the massive genocide of animals, the mind has some fun by becoming even more alienated. How can we not find pleasure with something that looks like our piece of Eden, a Paradise that deep down in our soul we know it exists? There, in the place that John Lennon eternalized so well in the song "Imagine" is the true tessiture of the real: without possessions, without hunger, without boundaries, without countries, without religions, with no future or past, just a "brotherhood of (wo)men/all". A place that today we intimately feel getting closer through the touch of screens that allow for us an instantaneous supra-territorial interaction in which everything is shared, liked, and commented on better known as "Facebook." Suffering like a dog in the crude reality of a body susceptible to an unjust and cruel world, it is no coincidence that said platform incites such collective fascination: in it we experience a mere inkling of that secret hideout of adventure that one day, when we were kids, we knew for sure to exist.

Where did we lose our certainties? In what moment we simply stopped believing? Why did we have to lose our faith? There is no greater pain than this: the pain of losing our innocence. Seduced by the Tree of Knowledge that gave human beings the power to judge Good and Bad, we paid the steep price of killing the purity of our child, deprived of its most authenticity. A necessary death. In order to survive in the veils of a world of hidden innocences, we soon learned how to be miscreant adults, full of doubts, resented, "embarrassed" of our own sex, of our own nudity. And since eating the forbidden fruit was the reason for our fall, we decided to technically exterminate not only the guts but we also decided to mitigate another factor of guilt concerning our bestiality: the instinct for sex. To eat became a sin, and a sin of the flesh. Thus, we not only built castrated screens but we also sublimated the coitus into the mental orgasm of the touch through the shock of the image. At last, the redemption of what we thought was the cause of our imperfection. It seems ironic that at the same time that we have never talked so much and so explicitly about sex, people are having so little of it[3]. While the movie industry implodes due to complaints of rape and sexual abuse of renowned actresses by well-known actors and directors, we adhere even more to the creation of copies, of profile images that are no longer susceptible to the unforgivable and programmed vileness of the instincts.

The redemption of the original sin that expelled us from Eden ends up by paradoxically creating angelical models of the Beyond-Human that live in harmony with the maximum exacerbation of control mechanisms (Deleuze, 2008). Control of the control no longer over things, but mainly over light, photons, and electrons, over nuclear fission and over the electromagnetic spectrum's tiniest forces in favor of strategies of enlightenment and electrocution of the body that aim to enlarge the mind through the blind hypertrophy of the vision. This is the case because, after having at all costs drained the forces of sexuality and turning them into screens, we will happily walk toward the annihilation of the last and most basic of all instincts: fear. Without any fear of the consequences of being closer and closer to the Tree of Life surrounded by an army of Angels armed to the teeth, we pretend to animate the things we so roughly create. And it is obvious that, unsurprisingly, they become animated! Ignorant as we are, we announce to have invented the wheel instead of realizing that we had simply manipulated something that is already inherently alive: not only the "I," but everything, even the tiniest particle of the Universe, is also a particle of the Infinite Consciousness! Iron, sand, earth, water, and all the inanimate elements, even the ones without nerves, can express their instincts by "enlarging and shrinking the inter-atomic and inter-molecular spaces" (Ánandamúrti, 2005, p. 89). However, blindfolded by the arrogance of thinking that we have the power to grant life to what seems to be dead, today we are moving closer and closer to the edge of the image's abyss: the resurrected body of the hologram.

Hungry for conquering the last barrier that still scares us, we envy Christ by pointing all the atomic bombs toward the final destination of Beyond-Humanity: the end of death in order to exist as image. Stupidly incapable of understanding that everything that is alive one day will die so that it can reborn, we connect, unify, and make everything omnipresent at the expense of a planet made of limited resources. All of this so that we can be free. Free from the animal limitations of a body tempted by

[3] Complete article at: https://www.vice.com/pt_br/article/bmbjwz/millennials-nao-transam-estudo, accessed on 07/27/2020.

advertisements to omnipresently inhabit the mind's holographic reality. "Ubiquity," a big word that is related to the property of something that is or exists at the same time everywhere, in everything, and in everybody or, in a nutshell, of something that divinely becomes omnipresent beyond the cycles of life and death. It is no accident that, in the science fiction masterpiece called "Ubik," Philip K. Dick tells us about a dystopian future controlled by corporative monopolies, authoritarian regimes, and altered states of consciousness in which dead people are treated like "half-lives" through maintenance systems of cerebral functions that allow for communication after death. Would this be a scenario produced by the convergence of mediation and meditation devices into a cosmic super-device whose "meta-program is beginning to program itself"? (Flusser, 2008, p. 79). Facebook is already announcing this desire for a cemetery of the living dead: dead people with profiles would defy another biological cycle thanks to a complex algorithmic, electrical, and human agency. Beings who, on the trail of a body that is gone and in the permanence of algorithms that are updated as a computer programming language, create subjects whose minds survive and no longer take their sustenance from oxygen and food, but from the power of electricity transmuted into an image on the screen.

However, long time ago Flusser (2008) has issued an alert: "the vision I propose, in which the objective world retrocedes and shrinks, and in which upcoming people will become increasingly fixated about oneiric terminals is, admittedly, a terminal vision of humanity" (p. 139). And the symptoms of this ending are being manifested right under our noses: unconsciously frightened by the consequences of the end of the instinct of fear, around 350 million of photos are daily published on Facebook[4] so that one day we can exist, that is, be remembered when we, by our own ignorant will, kill ourselves. Until then, we can be deluded by chanting solemn hymns to progress, shouting about the elaboration of symmetric anthropologies and object-oriented ontologies that equal humans to non-humans and reify everything into objects while forgetting one single detail: that "non-human" belongs to nature as much as we, human beings, do. After all, everything is nature. We only have to think about the geologic time scale to remember that we are the youngest sibling in this long process of creation and we don't even need to exist in order to our

[4] More information at: https://www.brandwatch.com/blog/47-facebook-statistics/, accessed on 07/27/2020.

much-mistreated planet exist as a living planet. Before the radicality of living, we are dispensable. Do we really want to pay the price of death to become, at most, an artlessly Angelus Novus? Isn't it time to think about what would be our "human nature"?

It is much easier to adopt a fatalist discourse and simply fold our arms when facing something considered to be "inevitable." But let us be optimistic, that is, let us think about fissures, cracks, ruptures, possibilities of opening when staring at the system's programmed closing. This is the calling that Flusser (1985) incites us to answer: when we alienatedly march toward an absurd death, it is even more urgent the appeal to the creation of a philosophy of photography as the "only revolution still possible" (p. 41). And what kind of revolution is this? It is a revolution that necessarily pertains to a revolution of the affections, of the consciousness, and of the most obscure corners of the unconscious mind: a revolution that is also a revolution of the desire. After all, what do we desire? What in fact do we want? How does the knowledge generated in the production of image devices can serve as a compass for our process of self-knowledge? If we consider the current stage of technical evolution as a social expression of our consciousness levels, it becomes easy to diagnose that today we are at the threshold between *manomaya* and *atimanasa kos'a*. How can we explain in any other way the fascination caused by Facebook from the possibility of instant communication, transforming each post into a work of art and globally connecting our minds? We want to acquire the creative insights, the premonitory dreams, and the paranormal powers of telepathy and clairvoyance inscribed into the plane of *atimanasa*. It is no coincidence that, we are gently surrendering ourselves with the use of an acceleration platform for the burning of *samskaras* through the unconscious interplay between actions and reactions posted by the profile's mind.

Nevertheless, the passageway toward the plane of *atimanasa kos'a* has in store a merciless trap: in it, not only *samskaras* are deflagrated, but we also find the snares of *aham*, the ego that, when claiming for itself the authorship of the action, takes the risk of considering such skills as "superpowers" and multiply even more *samskaras*. That is the necessary death we are desperately seeking: the extinction of our ego, but a death by its complete integration. And, between you and me, we all know that Facebook—or Instagram or TikTok—is a materialization of the narcissistic apex of egocentrism trying only to be mercifully accepted, socially validated. Would this be the reason why Flusser stubbornly talked about the necessity of the ascension of a "new consciousness" (Flusser, 2008)? One thing is certain: it is no use trying to cover the Sun with our fingers and being satisfied with the tip of the iceberg. We need as soon as possible to dive into the depths of the mind's screens, bring to surface the most putrid corners of the deep web[5], and identify in the current obsession for the image an ardent desire to, at last, rediscover the soul of our muzzled child before we fall into the abyss of self-destruction. We literally need to "get real." Stop wanting to abstract, levitate, get enlightened, confuse phones with people and people with phones, and mix-up the molecular level of life with its quantic dimension. All the energetic universe of Physics is not only attractive but also very interesting. However, clearly this is not the reality that we inhabit. Even I admit my complete cockeyedness when I thought I had found in the concept of "energetic of the existence" (Mozzini, 2017) the explanation for everything. Some years ago, I was talking about "deanthropologies" in such an abstract way that I am not sure if today I could understand what I was actually trying to say. All this effort to run away from myself and not face what Klee's angel stares, and, at the same time, is intently moving away from: the pain of trauma.

It is the damning pain felt in the flesh that pushes us toward the suicidal desire of converting our body into bodies of light. Bodies of light, that is, bodies on a higher level emancipated from "their mammalian condition" (Flusser, 2008, p. 127) because would not have glands, nerves, and hormones secretion that force us into the loneliness of feeling. No one can feel for us. No one can alleviate the anger, the sorrow, the fear, the anguish, the mourning, the sadness, and the bitterness that only we can truly know how deep they are buried in our chest. To feel hurts. That is the reason why, at all costs, we are becoming anesthetized and seeking strategies to desensitize the corporeal substratum through the simple desire to no longer suffer: we mediate, meditate, get numb, get high. We want to be happy, goddammit! Wanting to be happy "also is an innate instinct" (Ánandamúrti, 2005, p. 81)! But how can we be happy in a cruel world that sordidly forces us to lose hope? How can we enjoy happiness if we are condemned to hide the innocence of our eternal child? It is no wonder

[5] *Deep web* are websites not indexed by search engines or that can only be found through advanced encryption systems. In contrast to the so-called surface web, that is, the web usually employed by most people, the deep web is the category that has the biggest growth information-wise. There is no way to measure it, but it is estimated that the deep web can be at least five hundred times bigger than the surface web, sheltering all kinds of information and data, specially illegal ones, such as child pornography, drug-trafficking rings, criminal networks, and computer virus distribution networks.

that today we worship cats and dogs: they park their consciousness in the naivety of childhood. But us, humans, do not. We grow up and accumulate in our unconscious the most atrocious pains spitted out in the world by the mind's ghosts: are our technologies and social media a mere product of it as well? How is it possible for that helpless baby became part of a miserable, petty, mediocre, assassin, perverted, and corruptible "human race"? And since we no longer stand inhabiting the space of our own body infested with suffering, we become dissociated from it while hoping that, by throwing ourselves mentally into the image's abyss, we will find a way out of the death by loneliness.

But no. Besides the tragic possibility of our own extinction, there is no other option: we are doomed to feel. And it was like this, twisting in pain, that I found myself forced to live every word that was written here. Forced because, since March 17 of 2017, out of the blue I had a terrible panic attack right in front of my computer's screen while I was working on this book still in Valencia. Since the panic tends to repeat itself when someone faces what originally triggered it, there was no other choice but to deal with that ineffable pain: the pain of a sudden death[6]. What in me was hurting so much? Why something like that was affecting someone like me, who thought to be so "balanced"? I did not identify in my own madness the collective madness of desiring to be God. And facing the neglect of the doctor who was tending me at home and treated my despair as some sort of seizure curable by three days of prescription drugs, it took me a long time to finally realize that which had become a part of my life. In May, I moved to Berlin without treating myself and the panic, which was already bad, became chronic. While my distant friends and relatives would like my pictures posted on Facebook exalting the fact that I looked like a "citizen of the world," outside social media, my body was painfully suffering. I did not understand why I was nothing like my image and blamed myself for not feeling what I was impelled to feel and show as happiness.

[6] Panic attacks happen like a heart attack: out of the blue. And among its many symptoms we can find: feeling of imminent danger, fear of losing control, feelings of indifference, feeling of being out of reality, numbing and tingling in the hands, feet or face, palpitations, accelerated cardiac rhythm and tachycardia, sweating, shakings and spasms, difficulty to breath, shortness of breath, feeling suffocated, hyperventilation, shivering, heat waves, nausea, abdominal pain, chest pain and discomfort, headaches, dizziness, fainting, feeling of having the throat getting shut, and difficulty to swallow. However, without the shadow of a doubt, the worst one among all of these symptoms was the clear feeling of death or of an imminent tragedy.

Two people, in particular, were able to see through my smile: Larissa, my roommate, and Anita, from the Vilém Flusser Archive. Until that point, I thought my Panic Syndrome was related to finding the "right way," the "gist" of the writing. In July, Larissa saw me on the street while I was having a panic attack and, after gently accompanying me to the apartment, we had a conversation that finally made me understand that the problem was not my writing style, but that I needed to treat myself in order to be able to actually write. Anita was also fundamental. With surgical precision, she gave me three advices: fuck the deadline, postpone your Ph.D.; go to Australia because Ajit is real; and use your emotions to write. I was reluctant to follow all of her words and I did not comprehend what she was trying to say. But by the end of August I moved to Australia, in September I got married and, unwillingly, in December I had to postpone my deadline.

Once again, my world was turning upside down: to the writing's anguish was added the drastic reconfiguration of a married life in a new country where I did not know absolutely anybody or anything. Without money, friends, or relatives, without my language and in panic, I saw all my beloved ones becoming images on social media while I felt completely isolated, dependent, and vulnerable. And it was like this, losing control of everything I thought I had control over that, in the most harsh way, I had to undress my own castle of concepts: three years ago, I was researching the affections of ubiquitous image based on performing arts. More specifically, I was using the performances "Boxing" (1977), by Ion Grigorescu; "Theme Song" (1973), by Vito Acconci; "Open my Glade" (2000), by Pipilotti Rist. However, blocked by the Frankenstein of abstraction I created, in the last six months of my Ph.D. I had to start the writing all over from scratch, keeping what was most basic to my heart: mediation and meditation. Believe me: if I simply did not write the way I did horrendous panic attacks would just come out of nothing. The irony is that, after all, that pain ended up being my guide all along this journey.

Then, since I accepted medi(t)ation to become my research, I kept diving into a new and immense territory of unknowing that, even though I was no longer talking about the field of performing arts, I actually was: I had to perform the very act of writing. To perform because in a very raw way I lived, in real time, each line of this birth of words: leaving no stone unturned, I found in myself my worst enemy when facing the emptiness of the blank page to, in text, rip myself apart. It hurts to give birth to our own monstrosity. But when we face death, we are obliged to speak up and stop rounding. An infinite number of times I stopped, fell, quit, despaired, tripped, cursed, and started over with no truce from the first to the last minute of writing. It wasn't beautiful and "light." And after the writing finished, a new stage of omnipresence started: I had to go through an enormous bureaucratic process in order to have my thesis submitted to both the Polytechnic University of Valencia and the State university of Rio de Janeiro. Normally, this procedure takes five months to be done, but due to my deadline, I only had a month and a half. To get through this final deadline, I had to became present, from Australia, both in Spain and Brazil at the same time. During this whole period, my sleeping schedule just went to hell and I witnessed the effects of being omnipresent in my own body: it is just not humanly possible.

However, day after day, by resuscitating the rest of the instincts left in me I was able to survive this intense immersion process in the social unconscious through the screen of Facebook, and also in the individual unconscious through the screens of the mind, with a different perspective: it is no accident that today our smartphones are devices that we unlock so that we can access them. There, in that piece of screen through which we spit out our own thoughts via social media profiles, we also scroll the uneasy portion of our unconscious that we avoid digesting due to the simple fact that it's too painful. But what else can we do with the pain but to live it? Mediating and meditating, living the writing and being healed by it, I found in the production of knowledge a method of self-knowledge. Today, I realize that the feeling of panic that assailed me was also a way of bringing me back to my body and stopping me from dissociating. Dissociation is a psychic disorder[7] currently quite common in which one experiences the clear separation between body and mind; it is like being here but with your head elsewhere so to escape the present. And what is Facebook if not a technical stronghold for dissociated minds? Flusser (2008) was already prophetically saying that the universe of technical images implies the ultimate and "final negation: the negation of the objective world" (p. 138). By isolating ourselves into bubbles that temporarily help us to withstand the pain, we double our existence as profile images that "will create a magic circle around society" (p. 33).

How can we translate the real meaning of this circle that Flusser since long ago was seeing? My guess: we are going through an intense process of

[7] More information at: http://www.psicosite.com.br/tra/sod/dissociativo.htm, accessed on 07/27/2020.

programmed elevation and neutralization of the affections that employs discourses of spiritual enlightenment to promote the circuit's closing through the complete derealization of the real. Let us not get bewitched by such spells of the current General Economy of Light. But what can we do when facing the promising success of its program? How can we deal with the opportunistic proliferation of meditation Apps? Meditation is something to be taken with honor and respect. And also reliability: it is time for the tantric science to be more than just "something mystical." Together with psychoanalysis and neuroscience, the tantric perspective is vital for our comprehension of the human mind and its desires. Its systematization is the result of an inmesurable rich empirical knowledge that requires self-responsibility and, above all, a good antidote against the trap of distorting our innate desire for limitlessness into a desire for omnipresence.

That is why, when approaching the *debut* of this social media that in 2019 celebrated fifteen years of existence, it is time for us to honestly ask ourselves: why is it increasingly unbearable to inhabit the space occupied by our own bodies to the point that we became addicted to drugs like Facebook or obsessed with spiritual enlightenment? What are we denying, after all? It is urgent for us to reconcile the space of the mind with the space of the body so that we become what we truly are: human beings. Not Angels. Not Gods. Not Demons. We are fallible, dual, divine, and carnal, all of that at the same time. We are between heaven and earth, wanderers in the paradise that we inconsequently dare to destroy through the evil mistake of thinking that we can build it better than the Creator Itself. Let us no longer be deluded with Internet promises of omnipresence, of destitution of each and every remnant of time and space, and of actualization of the possibility of a Beyond-Human existence. To be human is to be present. To be human is to not know. Flusser (2008) wisely used to say that the problem of freedom "is not related to the ability of being opposed to this or that determination, but the ability to ignore all conditions in order to elaborate a non-determined universe" (p. 139). It is up to us to decide: until what point are we capable of standing such uncertainty? Until what point can we stand to decelerate, stop expanding, and let go of the reins of hypercontrol? Until what point are we willing to accept the "blind spot" of death that puts limits to our imagination?

If there is a limit, it is the limit of the body. Body-temple, first and last dwelling in this journey that we will only comprehend at the very end, when we all die. Body that inhabits the body of a Blue Planet better known as "Earth." Body that is our stronghold of self-knowledge. Ah! If we could unite the expansions of mediation and meditation, mind and screen, fun and discipline so that we could never again forget the mystery of who we are, we would no longer have such a traumatized world to live in. That is the reason why, if there is a model to be followed, it necessarily must be a model that allows us to keep our inner child, our innocence, our faith, alive. It was thanks to this child that, even in the dark, I could find the light I believed to be only at the end of the tunnel. The light is here, now, in the feeling of being present in one's own body, even though there is pain, fear, hunger, laziness, anger, but above all: love. Love is the only truly omnipresent thing that permeates this universe. And it was through the power of love, and from the imperfect human love between my partner and I, that even though I was struggling with shadows, I was able to find the balm that allowed me to keep going day after day. In August 2017, I traveled 29 hours by plane in order to cross the oceans that separated us. And I would do it a million times again. We met out of the blue, mediating and meditating, just like this book. But both the screen and the mind have a limit. The limit of the touch. We must take risks, take a leap of faith, throw ourselves into the unknown in order to find what truly makes us real. And in a world that incites us to the excess of accesses, maybe our biggest opportunity to resist is neither to dissolve nor to reify the border, but simply to inhabit it. Therefore, let us grab more the skin and embrace the micropolitics of affection: it is by feeling eye to eye, touch to touch, scent to scent, talk to talk, silence to silence that, even within our own limits, we can become limitlessness. And that in this journey we may find in the never-silenced question made by Espinosa the only doubt capable of being lived: after all, what can a body do?

References

Ánandamúrti, S. S. (2005). *Psicología del Yoga*. Valencia: Ananda Marga Publicaciones.

Benjamin, Walter. (1994). Teses sobre o conceito de história. In: *Walter Benjamin—Obras escolhidas*. Vol. 1. Magia e técnica, arte e política. Ensaios sobre literatura e história da cultura. Sérgio Paulo Rouanet (trad.). São Paulo: Brasiliense.

Deleuze, Gilles. (2008). *Conversações*. Tradução Peter Pál Pelbart. 7 ed. São Paulo: Ed. 34.

Flusser, V. (1985). *Filosofia da caixa preta: ensaios para uma filosofia da fotografia*. São Paulo: Annablume.

Flusser, V. (2008). *O Universo das Imagens Técnicas—Elogio da Superficialidade*. São Paulo: Annablume.

Flusser, V. (2015). *Comunicologia—Reflexões sobre o futuro*. São Paulo: Martins Fontes.

Mozzini, C. (2017). La sobrevida de la imagen: esbozos para una energética de la existencia. In: III Congreso Internacional de Investigación en Artes Visuales, ANIAV 2017—GLOCAL, 2017. In *Libro de Actas—III Congreso Internacional de Investigación en Artes Visuales: ANIAV 2017, GLOCAL*. Valencia: Universitat Politècnica València.

Vaz, Paulo. (2010). Esperança e excesso. In: André Parente (Org.), Tramas da rede: novas dimensões filosóficas, estéticas e políticas da comunicação. Porto Alegre: Sulina, pp. 189-208.

Appendix: Flusser's Letter to Dora F. Da Silva on the Phenomenology of the Body

Vilém Flusser

Post box 2, 49,590 Fontevraud

My dear Dora, I have received the following: your letter about letters, a copy of Augustine, Blue Horse 7, and the letter from Viki talking about the choir at the opening of the chapel. In short: letters are not actually "dialogical channels" in the sense of Aldo Testa, but channels of cross speeches made in Bagolini's "*otium.*"

The important are the intersection points. This one, for example: your letters have the same effect on me as mine have on you (a frustrated vital exchange), but consider the following intersection: You are engaged in a chapel for the worship of a Saint (Francis, I suppose), in disagreement with your time (that of seeking to eliminate all "mediation" in religiosity as in all fields), but in accordance with your space (that of mediumistic "mediations" and others).

In fact, it was proved that the cult of the Saints is a Celtic and Slavic tradition, not Greek and Latin, and not Germanic, and the attempt to abolish them, although framed in the current "ecumenism," it turns not only to the Gospel, but also to the Stoa. Well, I find myself in the same situation: what concerns me today, in disagreement with the Zeitgeist, is the mediation that my body exercises between me and my world. The cult of Saints is, in a sense, the introduction of corporeality between me and what is entirely different. "*Abschirmung*" is the term, do you understand it? The Saint is a "foreign body," but similar to mine, that I use, to protect

C. Mozzini-Alister, *Does Social Media Have Limits?*,
https://doi.org/10.1007/978-3-030-74120-4

myself, between me and the unbearable Divinity. Because direct light is dazzling. Under this approach, consider the following:

I was instigated, by a professor of "philosophy of biology," to contemplate the phenomenological matter of "organisms." What is biology concerned with, phenomenologically speaking? Certainly not with molecules like DNA, nor with the ecological balance of air and liquid solutions in the liver of a fish. Such "organismic" provocation came at a time when my own organism is problematized on two levels: it does not work perfectly, and provides me the delights of French cuisine and spring. You know my ambivalence regarding my body, and how it links to the Christian idea (for me indigestible), of the historical embodiment of the Abyss. In short: what does it mean to "have a body" (to be part of nature and history), to me, to living beings (and your Saints), and to Him? I would like to engage you in some of my thoughts about this.

I propose the following model: a pulsating "blastocoel" type ball (e.g., a snail in a shell, as you see it when eating *escargots*). At the center of the ball there is me (or "soul," or "nothing"). Around me are partially reversible eccentric and concentric arrows ("my body"). Around this, there are problems that are approaching, and products that are expelled ("my world"). Around this, forming a horizon, is my death (or; again, "the nothing").

But I am not only in the center of the ball, but also outside of it, which allows me to propose this model to you. And you are not only one of the problems in my world and one of the products I eliminate, but you are also above it, since I can propose the model to you. Our mutual "transcendence of the ball" is GOD, and I agree with Testa (and Talmud), on this point. Let us set this aside. The interesting thing now is to refine this model.

My body: this highly organized set of afferent and efferent arrows, however, confused. It is confused on this side with me, and on the other side with the world. For example: my "eyes" are close to me, because I cannot look directly at them. It's almost me. My "liver" is very far from me because I know it only as a "problem," that is, as part of my world. However: I can live without an eye, but not without a liver (although I know this not from experience, but from theory).

My finger is a concentric arrow (it brings problems from the world to me), eccentric, (solving problems and eliminating them from me and the world), but it is also me (while I write this letter). My hunger seems eccentric, and my sex concentric, but this will not withstand analysis (I am hungry to change the world and sex, so to give myself to the world).

Furthermore, is hunger and sex part of the body as is stomach and phallus, or is stomach and phallus theorized hunger and sex? But what theoretical entity is the stomach, if I can see the other's stomach and feel mine? And then: everything in my body may be hunger, including my eyes and my sex. And everything may be eye. Finally: I can deliberately transform my whole body into a problem (a worldly thing), as science of nature does itself, and I can deliberately transform my entire body into an eliminable product (as yoga does). Total confusion.

Of course: I have a world because I have a body. The body not only brings the world here, it precedes it logically and historically. It is because of body that I am in the world (*In-der-Welt-Sein*), and it is thanks to it that I have a future (problems to solve), and a past (problems that have been eliminated). In other words: "world" is my body's morphology, and "time" is my body's metabolism.

Which means that thanks to the body I am "immanent and mortal," that is, forward, valid, lucid, and inserted in the context. But there is more: since the body mediates between me and the world, it not only "produces" the world, but also shapes it. The world is as my body creates it. It can be very complicated.

First my body creates color, then my body manipulates problems, and then my body creates the electromagneticwaves that "are" the color in the background. And yet: my own body has colors, and in the background is the set of electromagneticwaves (a "field"). For, if my body is the model of my world, if everything in my world has my body as a model, the very model that I am proposing to you has my body as a model. Consider this, Dora: my body model has my body as a model. A melancholic result of the attempt to reflexively overcome corporeality.

My body is involved in everything. Maybe my liver less than my eye, but I'm not even sure of that. Every analysis in the world will ultimately reveal my body. And the analysis of "me"? Perhaps there is no last instance in such an analysis, because perhaps I am a hole in the center of my body. Perhaps this is why psychoanalysis is cretinism, or ("sin"), because "I" am "nothing" (a dialectical negation of the position of "my body").

Here our epistemological discourses intersect. Your "Saints," even, alas, Jesus, may be how our bodies shape the horizon of our world. On the one hand, bodies. On the other, negations of the "horizon" position, or "my death." There is a contradiction between "Saint" and "Francis" and between "Jesus" and "Christ." And the latter, that between "Jesus" and "Christ," is perhaps the Western dialectic between "me" and "death." It is urgent to carry a phenomenology of the body, before doing any type of Christology.

By the way: I can clearly observe the passage from the Romantic to the Gothic: the passage from the Pantocrator to the crucifix. The Pantocrator is not a body, it is the yolk of an egg. The crucifix is the body-cadaver. The Alpha-omega of the body? But then, if we want to be committed to History, the following problem arises: the Romanesque Pantocrator points to Saint James of Compostela and heavenly Jerusalem (VERACITY). I believe that Santiago is not a body (I never saw it, not even in the city of Comp Compostela itself, and you need "VERACITY"). As for Jerusalem, Agnon (the real Zionist), says that it has no body.

In fact, if you are a Zionist, you should know that Jerusalem is not *Hierosolyma* (i.e., paganism), but *Yerushalayim* = *IR BA DAVID KHANA* (the one who prostrated David). This is VERACITY. Such acid is the essence of the Romanesque vision of the body. But the crucifix is something else. It points to Maria, Notre Dame de Paris and Chartres, therefore the womb. The hole in the body, the vessel of sin, Maria is the male chauvinism, the "*Benedicta in mulieribus,*" which is the mediation between the males "father" and "son."

It is the VERACITY of Chartres. But what foreign body is Maria? Will the belly become a body? Look at Rachel, the prototype of Maria much more than the *Magna Mater* of the Gauls. The angel says to Rachel: you are not Leah, and within nine months you will be laughing. What paunchy Annunciation is this? Probably before all Jerusalemic VERACITY, a *Bethlehemic* VERACITY (the city of Rachel, after all). Before all Zionism a *Bethlehemism*; (Beth Lehem—House of Bread), since it is bread and wine that Rachel offers to the angel. And don't forget Isaac (*Yitzchak* = will laugh), in this. In other words: before all theology, a phenomenology of the body is urgent.

Dear Dora: Make our letters intersect. The cross, it is true, is a very ambivalent symbol. It means, among other things, "more," but also "died." But it also means "bivial" (decisive meeting of paths). Let us be bivial, not trivial (orthodox cross). Let us stay on the Maltese cross. Letters serve this purpose. And never write this terrible sentence again: I almost have the impression that I will never write again ... letters. Keep writing letters. Whether about Hoelderlin or VERACITY, Portugal or Jerusalem, there will always be crossing points. Because in certain ways we are moving toward the mutual encounter, and in another sense we move together. Feel embraced.

Reference

Flusser, Vilém. (05/28/1974), Unpublished correspondence. Nina Petek (translation from original letter in Portuguese). Vilém Flusser Archiv, Berlin (reference number: CORRESP. 9. 6-DORA (3126).

Afterword

To live is to accept oneself in order to chance oneself.

He who does not assumes himself does not live his own life, but the life ofpeople. He who assumes himself and accepts himself without at least trying to change does not live actively, but just functions in the function of what determines him. Because the attempt to change myself implies the attempt to change the surroundings in which I find myself. In short, to live is to discover who I am and to try and start from there in order to "be better"

(or "more"), thus changing mot merely oneself

but also the world.

Vilém Flusser

The process of writing this text has ended in June 2018. More than three years have gone by. I remember that in the day I finished it, I went to the quiet beaches of Frankston for a walk. I wanted to put my feet on the sand, feel the salty water of the ocean and give thanks to all the forces of nature that had supported me along this journey. I sat over my bag on the sand and stayed there, for a few hours appreciating the moment that I thought would never come. After feeling a mixture of exhaustion and gratitude, I decided to get a hold of my smartphone to take a photo of the sunset in front of me. However, for my surprise, the phone had just broken. The screen was completely in pieces, out of work. When I sat over my bag, I didn't realize I was actually sitting over the phone. Maybe it could no longer stand the weight of my existence. Or neither I couldn't stand it.

C. Mozzini-Alister, *Does Social Media Have Limits?*,
https://doi.org/10.1007/978-3-030-74120-4

I remember that I just laughed on such an irony: on the precise day I finished the writing, my phone was gone! I felt a relief. I felt freedom.

Since then, I just bought a new dumb phone and moved on with my life. Ajit and I were going to Brazil for three months to defend this thesis before the final examination board as well as for our Brazilian wedding. When we arrived at the airport, my father picked us up. Straight away, he gave me a smartphone. I silently gave a deep sigh and thanked him for the gift anyway. I strongly avoided using the device that was the cause of my addiction. However, with the arranging preparations of the wedding arriving, it ended up being inevitable. I was having double the work without that so undesirable object. So there I was again, on the screen and on social media! The Ph.D. presentation went very well, as well as the wedding. Ajit and I were just enjoying the moment and had no idea of what do to with our lives after. The only thing we knew for sure is that we had to be back in Australia on October for Ajit's Brother wedding. Our dream was to go traveling for a while and rest a bit—but reality just crushed us hard: we didn't have enough money, we quickly had to decide where to live, and we had to urgently find work.

After six months living in Brisbane, in the beginning of 2019 we went to live in the Sunshine Coast, close to Ajit's Family and to the University of the Sunshine Coast (USC), where Ajit would begin his studies on Education and Outdoor Education and I would start to work as an Adjunct Research Fellow. Only then I could slowly approach this text again to review and send it to be published. I was afraid of feeling panic again. Thankfully, with time I became free from that horrible pain. However, things were still very unstable. Since we came back to Australia, in less than a year and a half, we moved place eight times, living in shared houses with all sorts of people. All this transition was actually very stressful for the both of us. Ajit was moving from a very stable full-time worker life to a student life. I was moving over and over again since 2016 without a home to set my roots. At the beginning, I really struggled to adapt to the Australian society and several cultural shocks assailed me. It was as if, each single day, I would wake up with my feet on the roof and my head on the floor. I was feeling completely lost and in doubt if I should go back to Brazil.

These financial, transitional, and existential problems, of course, affected our relationship. At some point, the cultural differences were present between us. Ajit and I had to put an effort to overcome our own fears and get in touch with the love that brought us together in February 22nd, 2017. As I said before, we literally met mediating and meditating, just like this book. Out of nowhere, I received a Facebook message that I might had never seen because we weren't friends on social media. I was living in Spain, Ajit was in Australia. But one common friend from our meditation worldwide group was enough for him to reach out to me with the cupid's arrow: straight away, in that first message he said he was looking for someone "spiritual minded in order to build a relationship." Direct and frank, his message scared me. However, given the good references that our common friend gave me and the obvious evidence that he was a gorgeous man, I agreed to talk to him. Since then, we simply never stopped talking! We stayed like this, distance talking, for four and a half months. With no games, no fear, naturally and confidently. Then he came to visit me in Berlin in July 2017. We traveled together through Italy and Denmark and, after two weeks, he proposed while we were camping in the middle of a chamomile field. A month later, I arrived in Australia. And in another one, we were married—and with this arduous thesis to handle.

I am saying all this to deeply express my gratitude to my partner, Ajit Nicholas Mozzini-Alister. Without Ajit, this book probably wouldn't even exist the way it is. He was the one who encouraged me to let go of all the unnecessary abstraction and focus on mediation and meditation. He was the one who supported me writing in first person, something I was afraid of due to do my academic conditionings. He was the one who listened to all my ideas and helped me find the shortcuts to simplify the writing process. He was the one who read each line of this book: every day when I arrived home, I would put what I had done in Portuguese to Google Translate it into English and, if he couldn't understand what I wrote, I would rewrite until it was clear enough. Ajit was the best support I could ever had, and I truly needed his encouragement to be able to say what my heart desperately wanted to express. Words are not enough to describe how much I love this human being and how much I appreciate his presence in my life. Thankfully, our love resisted all the initial difficulties and was bigger than any barrier or cultural difference: what is the need for words when silence speaks for itself?

I must say that, during this time, I haven't stopped meditating. However, I see meditation beyond the strict practice of it. Meditation can also be a walk, a run, a dance, a book, a talk, a pray, a food, a sunbath, a long hug ... Anything that helps us align the innate desire of our minds to be unlimited with the simplicity of our hearts and the grounding of our feet. Meditation, therefore, became the practice of surrendering the need for control to turn our own lives into a "medium" of work for this planet and everything there is in it, including us, human fellows. I gave up on obsessively seek for samádhi as well. Samádhi is real, it does exist, but it is a "grace," something received without any kind of merit or deserving. And it was certainly by grace that, after so much instability, in February 2020 we found a cozy home to live only the two of us, in the Sunshine Coast. Finally, among the quietness of the Covid-19 pandemic and social isolation, I could ground myself and have time to fall in love with my life in Australia. Today, as I write these words laid down in the sofa that we have so happily chosen together, looking at the big world map frame on the wall and the trees in front of our balcony, I fell at peace. Do you know why? Because life gifted me not only with love, friends and a community, but also with a meaningful purpose.

All this time I was searching for the moment I would finally realize this choice. Some people are lucky (or blessed) enough to born knowing what they want to do since they were kids. That was not my case. Life for me was always a messy endless puzzle that I had to put the pieces together in order to see the big picture. Well, I believe the final piece is only found in the end, with death, and many things will change until there. However, at least this very special corner of it is assembled for now. What would happen if I transformed my passive and addictive use of social media into something propositive, disruptive, and meaningful? And what if other people could do the same? I realized that instead of cutting off social media out of my way, I had to hack it, shake it from the inside out. My goal is not only to facilitate a purposeful use of social media among people, but also to raise awareness to the not very inspiring sides of these platforms: digital traces and their use for marketing and political aims, polarization and social fragmentation, social media addiction, specially among kids and teenagers, anthropocene and the cost of our mediation technologies, and above all, the mental health effects of our screen media. We need to frankly discuss social media in order to educate not only our use of such social networks, but mainly to educate the social media companies themselves. Could our own desire for omnipresence also be the beginning of different ways of using social media? Would we have the power to reshape the infrastructure of our smartphones and social networks into something sustainable? I have the hope that if we start to at least demand, as users, the privacy of our personal data and fight for it as a vital cause, we might be able to stop the missuses of such information and begin to turn social media in favor of our entire humanity.

I say this because, if this does not become possible, then I am afraid that there is no other solution than to destroy social media before it destroys us. If we keep unconsciously moving driven by our thirst for progress for the sake of progress in order to realize the resurrected body of Christ, then we will be in real danger. Danger of totally converting the subtlety of our mental process into mere automated systems of energetic symbiosis with devices. Danger of living in a society of hyper control for the arrogance of not dealing with the mystery of death. Danger of trying to technically grasp the soul for the avoidance of having to deal with uncomfortable emotions. Danger of increasing so much labor exploitation and environmental destruction for the fear of facing the pain of trauma. Danger of not recognizing the limits of our own bodies for the vanity of wanting to become a grotesque version of God Itself. Do we really need to turn the resurrected body of Christ into a kind of anthropological destiny? Is this the future we want to give to our children?

As Flusser would agree, the current technologies are nothing but a result of our outer expansion, a mirror of our minds being externalized for us to know ourselves better. It is clear that we are desperately seeking for "enlightenment"—whatever that may be—through screens that are literally inserting "light" into our nervous system. We all want to shine bright just like stars which, unfortunately, are barely visible in nowadays polluted skies. Nevertheless, deep down, we still recall the presence of their cosmic-dust in us. Thus, if we could collectively stop to do an inner expansion exercise of self-knowledge, could we actually create different technologies? Would we also have space for the wisdom developed by Aboriginal technologies, Indigenous technologies, or African technologies that have been—or attempted to be—wiped out along "History"? My goal now is to address the core of our current addiction to social media and help society to consciously shape our hegemonic technologies as a tool for self-knowledge in four different levels: personal, social, environmental and humanitarian, as well as spiritual. Maybe, someday, we will come to the conclusion that we don't even need such technical mediation devices due to the fact that we already have all of its possibilities within us. I don't know.

However, what I do know is that, specially with the current pandemia of SARS-CoV-2, better known as "Covid-19," and the consequent increasing of our digital imprint, it is urgent to question the limits of social media and transform our current desire for omnipresence into an authentic "*desire for presence.*" How can we do that? Well, I'll leave this question open for my next book, which I hope to write no longer as a body of light: instead, I want to do it as a body of flesh.

25 June 2021, Buderim

Index[1]

[1] Note: Page numbers followed by 'n' refer to notes.

C. Mozzini-Alister, *Does Social Media Have Limits?*,
https://doi.org/10.1007/978-3-030-74120-4

D

E

Q

R

The manufacturer's authorised representative in the EU is Springer Nature Customer Service Centre GmbH, Europaplatz 3, 69115 Heidelberg, Germany. If you have any concerns regarding our products, please contact ProductSafety@springernature.com

Printed and bound by CPI Group (UK) Ltd, Croydon, CR0 4YY

22/09/2026

02221066-0001